SISTERS IN SCIENCE

SISTERS IN SCIENCE

Conversations with Black Women Scientists about Race, Gender, and Their Passion for Science

DIANN JORDAN

Purdue University Press
West Lafayette, Indiana

Printed in the United States of America.

ISBN 978-1-55753-386-9
1-55753-386-5

In loving memory of my aunt and second "mom," Mrs. Mary Frances Grooms Pratt; without her I would not be the woman that I am today. I will never forget how you influenced my life. *Rest in Peace.*

And to the memory of Dr. Jann Patrice Primus, a daughter of America. May her legacy and spirit transcend all racial and gender barriers in science.

Contents

Preface

In her book *Find Where the Wind Goes: Moments from My Life*, Dr. Mae Carol Jemison, the first black woman astronaut, relates a very revealing childhood experience that epitomizes the early challenge many potential young black female scientists face when selecting science as a career. When a teacher posed the question what she would like to be when she grew up Mae replied, "I want to be a scientist." Her teacher tried to convince her that what she meant was a nurse. Jemison, though only a child, was undaunted and persisted in stating that she wanted to be a scientist. In the 1960s, that was a pretty bold stance for a little black girl to not be persuaded by an authority figure. There are many potential Mae Jemisons in the world, but many of their dreams are actually never realized. How many goals are not met because some teacher, parent, or peer may have innocently dissuaded them? How many girls or young minorities become convinced that somehow they are not capable of doing math or science? Unfortunately, one of the earliest hurdles that females and minorities often face is similar to Jemison's story and that of other women scientists.

Research continues to show that girls are eager to study math and science, but around middle school girls tend to give up their science. For black girls, the dilemma of choosing a science career is compounded by issues of race as well as gender. While it is true that young black women actually go to college more often and earn more degrees than their black male counterparts, their numbers are a small percentage at the graduate level in some science and technology fields. A study by James Jay (1971) on black doctorates in America from 1876 to 1969 documents that only 58 of 650 doctorates were awarded to black women. Recent studies by the National Science Foundation (NSF) do not show a great deal of improvement for black women at the doctoral level in science, especially for those in mathematics and engineering (Jordan, 1997). For example, in 1995, 15 black females, 58 black males, 320 white females, and 1,766 white males received doctorates in engineering across the United States. Despite these dismal statistics, some black women, like Jemison, persevere to become highly successful scientists and engineers. The question then remains, Why do so few black women enter scientific careers, and for those who do, what can they teach us about how black women survive and succeed in science?

A number of significant factors contribute to black women's success in science. Lynda Jordan, a biochemist, has candidly discussed how race and gender have impacted her career as a student in a predominantly white institution as well as professionally in a historically black institution. While there is no denying that race and gender may affect success in science, they are not the only factors that contribute to black women's success and achievement. Familial support in childhood and adulthood can have a profound effect on black females entering and staying in science for the long haul. One need only look at the Mae Jemison story to see the power of a loving and nurturing family. A supportive spouse or a strong support system can also play a vital role in long-term success. Some of the women interviewed here are very candid about their marriages and how family life affected their career goals. For example, Freddie Dixon talks about how important it was to have a husband who did not mind cooking and taking care of their young child while she was in graduate school. Equally important but not always discussed, class can have a profound effect on the black female's desire to study science and the available opportunities to pursue it as a career. The late Jann Primus openly discusses how being from a privileged background influenced how she was viewed and treated as a scientist.

For too long the lives of black women scientists have been virtually invisible and often neglected in the larger American society and even in their own culture. My goal has been to let the black woman scientist speak for herself and tell her story in her own way. This goal, along with my lack of experience as an interviewer, may have influenced the flow and content of the interviews. I began this project over nine years ago with the intention of discussing three major areas of interest with the women. As I began to get to know them and discuss issues, more questions would arise, so that the later interviews may appear more detailed. Because I was a full-time practicing research scientist, my time spent on this project was limited to my spare and vacation time. Therefore, some of the interviews have greater content while others lack some needed discussion. Of course, this is also due in part to the interviewee's own comfort level in discussing matters of race, gender, and family life. Some of them were very forthcoming, while others preferred to not comment on specific issues related to race, gender, or family life. There were others who commented but were not comfortable allowing those comments to be published. A few times, I, as an unknown interviewer, may have been viewed with some suspicion, too. After all, little research has been done on this area of black women's lives. Far too often, the image of the black woman is not positive in print. Therefore, I, too, had to gain the trust of some of the women, even though I am black and tried to make clear my intention to present the black woman scientist in a positive light. Despite these inherent problems in the interviewing process, the writing of *Sisters in Science* has been one of the most rewarding endeavors of my life. The women who are presented here have wholeheartedly supported me and, in a few cases, allowed me to re-interview them to get the best story possible.

After a long, tiring day of doing research in the laboratory and the field, dealing with sometimes difficult colleagues, and engaging in a disproportionate share of university and community service, I still would come home and work on this project. I even used my holidays to gather and collect data on this group. I spent a great deal of time establishing a relationship with many of the women scientists. In fact, I loved doing the interviews. They made my own journey as a scientist much more bearable because I didn't feel alone as I talked with these women, my sisters in science. Consequently, when I sent the interviews back to the scientist, I had high expectations. Some of those expectations were disappointed when it sometimes took months to receive a response—and some never responded. I understood that these women, too, were inundated with all kinds of responsibilities in their everyday lives as scientists, but I wondered how I could tell their stories without their input. As I matured in the process, I learned to accept that a 50 percent return rate was better than no rate at all. I also had to accept my role as a full-time scientist and educator and that my limited time was a factor in doing the project the way that I may have desired. In a few cases, I simply did not return the transcribed interviews in a timely manner, which may have affected my response rate. Moreover, there were other outstanding black women scientists that I wanted to interview, but time simply did not permit me to do so. In any event, the process of doing research on the lives of black women scientists has been a labor of love that has opened me up to a different writing and critical thinking process as a scientist and the beginnings of an untold, fascinating history of American women.

When I was asked to do a presentation on black women scientists as a part of the Black Culture Center Week at the University of Missouri, Columbia, I thought, "Sure, that shouldn't be a problem. I'll just go to the library and get some key references and prepare for my presentation." I was shocked to find so little information on black women scientists; for that matter, there wasn't that much research on African American scientists—period. I slowly began that journey of trying to locate relevant sources. I even wrote and published a few articles myself. After my presentation on the subject in 1995 and receiving tenure and promotion in 1996, I decided that I had to tell our stories in a book format. Thus, *Sisters in Science* was conceived.

I still had to struggle to find time to interview black women scientists, collect data and resource material, and travel to the institutions where many black women get their first glimpse of science—historically black colleges and universities (HBCUs). I simply planned my vacations around that kind of travel. I was determined and passionate about finally telling some of our stories. Fortunately, most of the women were eager to talk and welcomed me with open arms. It was the beginning of a dream come true. Initially, I interviewed black female graduate students, who provided some information; but because students are transient and my schedule was tight in the early days, that process proved to be less productive. I think my overall goals were best served by interviewing the more established black woman scientist.

Although it may be somewhat of a challenge, I would encourage other researchers to consider studying this group and documenting their lives and stories. Over the years, I have also interviewed and talked with black women scientists from other countries who have particular challenges in the scientific and engineering professions, and their stories are worth investigating further. There is also a whole group of black women scientists who never pursued doctorates in science or engineering but are quite active in industry, government, and educational institutions in their chosen profession. I am fortunate to include two such women, Hattie Carwell and Yvonne Clark, in this volume. Their stories and research about them should be told in a much more comprehensive way, since they represent a large percentage of the practicing black women scientists. There is still much work to be done on the scholarship and stories of the black woman scientist. The harvest is plentiful but the laborers are few.

Acknowledgments

When I was nine years old my mother bought me a pair of "microscopic" binoculars. It was the closest thing that she could afford to a microscope. She wanted to nurture the "scientist within me" and encourage me to be whatever I dreamed. My mother has been one of my strongest supporters throughout my sometimes rocky road to a successful scientific career. She, along with other family members, has been my Rock of Gibraltar. I thank Elizabeth Grooms Jordan, my mother; Willie James Jordan, my father, and Shirley Marie Jordan, my sister, for constantly being there for me. I especially thank Shirley for the initial typing and editing of the manuscript and our lively debates over the issues of women and minorities in science. You have lifted, listened, and loved me when I felt like I could not go on. You are my sister-friend extraordinaire.

I have been blessed with a beautiful extended family—my second parents, Eddie Lee and Mae Willie Jordan; and siblings, Willie Jr., Charles, Leona, Clement, Aretha, Demetruis (deceased), Eddie, Jr, Dionne, and Sharon—who have been a great source of support. In the black community, no one person just "raises" you; you have a whole entourage of aunts and uncles, cousins, and grandparents. It took a village to raise me. I thank my aunts, Nasrine Smith, Mildred Ricks, Lovie Lee Mitchell; and my uncle, Joe Timothy Grooms, for being so encouraging of my academic pursuits, especially during my childhood.

There have been many great teachers, scientists, and community leaders who influenced me along the way. A few people stand out in my mind. My junior high school science teacher, the late Charles Dockery, gently nurtured my first signs of interest in science, along with my English teacher, Margaret Buford Jones, who simply gave me an outstanding background in literature and writing. Later, my tenth-grade biology teacher, Richard Carter (affectionately known as Coach Carter), really spurred me on by encouraging me to do my first scientific experiment and present it to all the students taking biology in my high school. Of course, my initial motives for taking on the extra assignment were to ensure that I got an A for the term, but he wanted to encourage and expose me to an audience early on. I was a nervous, skinny 15-year-old, but he assured me that I could do anything that I really wanted to do in my life. The experience helped to give me the confidence to go to Tuskegee University and successfully complete a program in biology. Later I had the good fortune to meet and work with some great soil scientists at Alabama A & M University. They encouraged and mentored me to become a soil microbiologist. At Michigan State University, I met my next two mentors, James M. Tiedje and Charles W. Rice. These two people were significant in my survival of a rigorous Ph.D. program, but more importantly, they have mentored me in my scientific career. In the soil sciences, the number of black women doctorates is fewer than the fingers on both hands, even in the twenty-first century. I thank "Jim" and "Chuck" for being good mentors, friends, and kind colleagues.

The kind acts of my early and latter mentors led me to a successful research career at the University of Missouri for over ten years. I thank all colleagues and friends who shared and contributed to my success during my tenure there. As I have restructured my life goals and aspirations, I have been fortunate to have met others who have recognized what I can contribute to an organization. After leaving the University of Missouri, I met T. Clifford Bibb at Alabama State University. I thank him for giving me the opportunity to began anew and share my talents and skills with the ASU family. My colleagues (Shuntele, Emma "E C," Kartz, Shree, and Gladstone in biology, and others) have welcomed me home with open arms in my new department.

I have met a lot of people over the years, and some have remained good friends. I have been friends with Judy V. Burton for 25 years. She has supported me wholeheartedly throughout my career and life. Thank you, Judy. You are a gem of a friend. A special thanks to Gregory McDonald for being a "good brother in science" and forming a research collaboration with me and my students when I needed it the most. I would be remiss if I did not mention my friend John McClendon, III, former director of the Black Culture Center at the University of Missouri-Columbia. John provided the first forum where students and faculty could begin to discuss the issues of race and gender in science with his seminar series. Those fruitful discussions challenged me to begin this project.

I would also like to thank the staff of the library and archives at Tuskegee University (Cynthia Wilson), Spelman College (Taronda Spencer), and Howard University (Joellen Elbashir) for providing assistance in locating information on and photographs of black women in science. James Henderson, emeritus director of the Carver Research Foundation and professor of biology at Tuskegee University, was kind enough to share what files he had on the subject. I am grateful to Dr. Henderson for his generosity and teaching me some environmental science. I thank Beverly Guy-Sheftall (Spelman College), Susan Hill (National Science Foundation), Jill Tietjen (past president of the Society of Women Engineers and consultant), and the late Barbara Lazarus (Carnegie Mellon University) for their support in this project. I never met Barbara Lazarus, but she always supported my work and so many other programs to increase women and minority participation in science. Most importantly, I thank all the scientists who participated in this project (published and unpublished), and especially those who stuck with me through the thick and thin of it all. I am deeply grateful to the family of the late Jann Patrice Primus, especially her sister, Jeanne Johnson, and mother, Ruth Boston Primus, who granted me permission to publish the interview and provided me with valuable insight into Jann's life and early years. I would also like to thank the director, Thomas Bacher, Margaret Hunt, Donna VanLeer, and Bryan Shaffer at Purdue University Press for believing in the project, being extremely patient, and providing me the opportunity to tell our stories.

And above all of these wonderful people who were brought into my life for a reason at the right time or place, I give all the glory to God.

INTRODUCTION

The composition of the workforce is steadily changing at the dawn of this new century. When my great-grandmother "Nettie Jay" was born near the end of the nineteenth century, most black women could only find employment as maids or field hands. Only 5 percent of black women could read and write, and the prospects of attending college were almost nil. Today, women and minorities are able to work and compete with the best and brightest in many professions, including science and technology. Most studies, however, show that women and minorities, despite their growing numbers, are still primarily working in the service-oriented and lower-paying jobs in the American workforce. Yet the numbers of black women in science, technology and engineering are growing and are certainly better than in 1894 when my great grandmother had virtually no chance of a science career as I have had.

Blacks make up about 12 percent of the American population, yet they still only represent about 2–3 percent of the scientific professionals (table 1). Often black women represent less than a single percent in their respective discipline, especially at the doctoral level. Though their numbers are miniscule in comparison to the number of whites in science, there are still more black women in science than there have been at any other time in our nation's history. Yet most people are not even aware of these women's contributions to science and technology. Much has been written about women in science, some on minorities in science, but very little has been written about black women and other women of color in science and technology. Their stories have not been told; their voices are not heard.

Sisters in Science allows the women to speak for themselves in response to questions conducted in an interview with the author. This approach allows the women to tell more of their own story in their own words. We learn of the early influences—family, school, geography—that shaped their decisions to pursue science and of their experiences as "students of science" in their climb through the academic maze and their ongoing research interests. As professionals, they also discuss the impact that factors such as race and gender have had on their lives as women and as scientists; they also reveal the degree to which the civil rights movement and the women's rights movement played a role in their success. Along those same lines, these scientists reflect on their continuing struggle to become visible in a white, male-dominated world, the future of women in science, and their perception of their contributions to science and technology. Perhaps most insightful is the women's views on balancing the roles of wife, mother, and scientist and the factors that have been most critical to their success in each role. So many young women are simply turned off by science when they realize the enormous personal sacrifices they have to make to conduct quality research and teach.

To understand the black woman scientist's place in history, a brief examination of African Americans and women in science must first be considered. I have tried to give a glimpse into the black woman scientist's story to show how dual identity as black and female has historically determined how she would define herself in America.

Brief History of African American Women in American Science

In order to understand the current plight of black women in the sciences and engineering, it is important to have a historical perspective of how both their race and gender impacted their development in science. Black women may have been involved in scientific investigations during the colonial period, but it is difficult to document their contributions due to their status in American society. Manning (1993) writes that slaves were known for their inventive abilities, but their legal status prevented them from holding patents and from achieving widespread public recognition for their achievements. Despite these limitations, the first black American was granted a patent on March 3, 1821 (Carter, 1989). Thomas L. Jenning, a tailor living in New York City, developed a method for dry cleaning clothes. About 64 years later, in 1885, the first patent was issued to a black woman, Sarah Goode, who patented a folding cabinet bed on July 14, 1885. These few exceptions somehow managed to achieve some level of recognition before and after the Civil War.

Research on black women's lives and their professions reveal that most black women were not privy to a formal education prior to the Civil War (Hine et al., 1993; Davis, 1982; Jordan; 1997; Thomas; 1989). Slavery and poverty were major deterrents to their obtaining even a meager education. During slavery, black women worked mainly on farms, as manual laborers and house servants (Brown, 1975). After Emancipation, black women still earned little or no wages for farm work and unskilled labor jobs. After the Civil War, the number of blacks (mainly men) entering the scientific disciplines slowly increased. This was due, in part, to the establishment of more black educational institutions. Before the Civil War, there were only three black colleges in the United States: Cheyney

State College (1839) and Lincoln University (1854), both in Pennsylvania, and Wilberforce University (1856), in Ohio (Clewell and Anderson, 1995). After the Civil War, more black institutions were established. According to Reynolds and Tietjen (2001), the path to education was even more difficult for minorities than for other segments in early America. If a white woman was perceived as being mentally incapable of absorbing the same education as a man, and non-whites were considered inferior to whites, then the plight of the black woman was essentially doomed. The whole idea of educating African American women before the Civil War was not even entertained.

Still with all the odds against them, black women slowly began to earn degrees. The path was forged by Mary Jane Patterson in 1862, when she distinguished herself as the first African American woman to earn a bachelor of science degree, in English from Oberlin College (Hine et al., 1993). In the late 1800s and the early 1900s a few black women had earned bachelor's degrees. By the 1920s they had progressed and successfully moved to the highest graduate degrees. In 1921 the first three black women earned Ph.D. degrees in economics, English, and German, though none were noted in science or engineering (Hine et al., 1993).

In science-related fields, such as medicine, there were a few exceptional cases. Because blacks were allowed to practice medicine, dentistry, law, and teaching in their communities, these professions appear to have been more accessible for African American women, and a few African American women earned degrees in medicine and dentistry in the mid- to late 1800s. In 1864 Rebecca Lee was the first black woman to become a physician in the U.S.—only 15 years after Elizabeth Blackwell became the first white woman to graduate from an American college (Hine et al., 1993; Davis, 1982). Ida Gray set up a dental practice in 1890 (Hine et al., 1993; Davis, 1982). By 1890, black women physicians made up about 2.6 percent (115) of all women physicians (Hine, 1993). These women received degrees from established black colleges, such as Meharry Medical College and Howard University, and northeastern, predominantly white institutions.

If black women had been practicing medicine since the mid-1800s, the question remains, "Where were black women in the scientific professions?" I believe that it can be inferred that they were obtaining a scientific education through the pre-medicine, medical, and dental curricula during this time as well. Close research shows that they, too, were gaining a small foothold in the scientific professions. In *Black Women Scientists in America*, Warren (1999) describes two of the earliest black women educated for scientific careers, Josephine A. Silone Yates and Beebe Lynk. Yates earned a degree from Rhode Island State Normal School in 1879. She later became a science professor at Lincoln University in Jefferson City, Missouri, and was the first woman to receive a full professorship and to become head of a department of natural sciences, in 1888. According to Warren (1999), Lynk earned her first degree from Lane College in Jackson, Tennessee, in 1892. In 1901, she studied pharmaceutical chemistry at the University of West Tennessee. In 1903, she earned a doctoral degree in pharmacy and became a professor of pharmacy and chemistry at West Tennessee. Clearly, a few black women were involved in science by the late 1800s and early 1900s as teachers and educators, but it would take 30 more years before they earned advanced science degrees.

By the early 1900s, some of the graduates of the newly established HBCUs and the handful of graduates with advanced degrees from predominantly white institutions were the shapers of the curriculum, and they provided direction for a future generation of scientists. Of particular note was Ernest Just, who received his Ph.D. from the University of Chicago in June 1916 (Yount, 1996). Just lectured and trained hundreds of students interested in science, medicine, dentistry, and health-related professions at Howard University. He also mentored Roger Arlinger Young, his research assistant at Howard University;

Oldest documented photograph of black women working in a science laboratory (Spelman College, 1919, Tapley Hall).

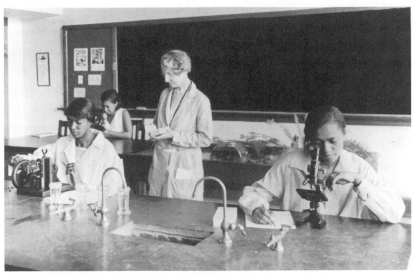

Second oldest photograph of black women in a science laboratory (biology) (Spelman College, circa 1922).

Young eventually became the first black woman to receive a Ph.D. in zoology from the University of Pennsylvania in 1940.

World War II brought the country together for a common cause and created many new opportunities for African Americans and women. Because white men and some black men were off to war, the doors opened for women and minorities with technical and scientific experience to contribute their expertise. African Americans, as a distinct group, began to gain some public attention for the first time. Some of these opportunities trickled down to a few African American women. In the 1930s and the 1940s, the first few African American women earned doctorates in the sciences. Ruth Moore earned a Ph.D. in bacteriology in 1933 from Ohio State University; Marguerite Thomas in geology in 1942 from Catholic University; and Euphemia Lofton Haynes in mathematics from Catholic University in 1943—with Evelyn Boyd Granville (Yale University) and Marjorie Lee Browne (University of Michigan) following Haynes's lead in 1949. Marie Daly earned a Ph.D. in chemistry in 1947 (Davis, 1982; Rossiter, 1995), and Jesse Jarue Mark received a Ph.D. in botany in 1935 from Iowa State University (Rossiter, 1995). Mark was the first African American to receive this degree, and she was the first black woman scientist on the faculty at a major white institution (Warren, 1999). Many other firsts are noted in the timeline in this book.

Although white women were ahead of black women in receiving degrees, they had, and still have, their own share of obstacles along the path to scientific achievement and equity. Until 1837, when Oberlin College began to accept anyone who sought higher education regardless of race or gender, not a single college in the United States admitted women (Reynolds and Tietjen, 2001). Oberlin was founded in 1833 as a seminary for men but later developed into a college. At first women only had a limited curriculum, but this lasted a short time, and a woman was graduated with a regular bachelor's degree in 1841. As for black women, Oberlin proved to be a good starting point. About 20 years after the first white woman graduated from Oberlin, Mary Jane Patterson received her bachelor's degree in English.

White women had been receiving advanced degrees in the sciences since the mid-1800s. Their experiences are documented elsewhere (Rossiter, 1982, 1995, etc.), but to place the black woman scientist's history in context, we shall briefly describe white women's accomplishments here. White women were involved in scientific investigations in horticulture, botany, and agronomy during the colonial period. As early as the 1700s, women were working in botany, with Jane Colden cataloguing more than 300 plants in the Lower Hudson Valley in 1757 (Heinemann, 1996). Apparently, botany and other disciplines related to plants seemed to be an acceptable scientific endeavor for women.

In other fields, Rachel Lloyd received a Ph.D. in chemistry in 1886 from the University of Zurich, and Florence Bascom was the first woman to receive a Ph.D. in geology, in 1893 from Johns Hopkins University (Heinemann, 1996). On the other hand, fields like engineering were very slow to allow women of any race into the professions. According to Reynolds and Tietjen (2001), Elizabeth Bragg became the first woman to obtain a civil engineering degree, from the University of California in 1876. She was followed by Elmina Wilson, who graduated in 1892 from Iowa State College (now University) with a civil engineering degree, and Bertha Lamme, who graduated in 1893 from Ohio State University with a degree in mechanical engineering.

Unlike their white female counterparts, African American women did not achieve their first engineering degrees until the 1940s. Black women, however, were receiving degrees and training in mathematics. For example, Blondelle Whaley received her M.S. degree in mathematics in 1929, while Hattie Scott received her B.S. degree in civil engineering only in 1946. Yvonne Y. Clark, featured in these interviews, was the first woman to receive a B.S. degree in mechanical engineering, from Howard University in 1952, four years after the mechanical engineering program was established at Howard. According to Clark (personal communications, 2002), women had already received bachelor's degrees in the other engineering programs at Howard University. White women received advanced engineering degrees considerably earlier than black women—one woman received a Ph.D. in engineering in 1920, and three women received Ph.D.s in the 1930s (Reynolds and Tietjen, 2001), but there were still no women on the engineering faculty at any of the 20 largest doctoral universities in the country in 1938. The situation was even more dire in the black community. The first African American male to receive an engineering Ph.D. was in 1925 in civil engineering (Jay, 1971). For African American women, it took over 50 years before they began to receive Ph.D. degrees in engineering, more than three decades after they had achieved B.S. degrees. Jennie Patrick (featured in this volume) received a Ph.D. in chemical engineering in 1979, and Christine Darden received her Ph.D. in mechanical engineering in 1983.

The Role of the Black College in Educating African American Scientists

After the Civil War and during Reconstruction, about 150 black colleges and institutions of higher education were established. Today fewer than 114 black colleges and universities still survive. Some of the private HBCUs were Spelman College in Atlanta, Morehouse College in Atlanta, Bennett College in North Carolina, Bethune Cookman in Florida, and Tuskegee Institute (now University) in Alabama. Tuskegee Institute was a unique institution of higher learning because, although it was considered a private institution, it was very much involved in the practical courses of agriculture and industrial arts. Some of the public institutions were Alabama A & M University, North Carolina A & T University, Lincoln University (Missouri), Florida A & M University, and South Carolina State University.

According to Trent and Hill (1994), these institutions are under constant scrutiny for their quality, and the need for their existence is often questioned. On the other side of that coin, although research and scholarship may not be the mainstay of all of these colleges, they provide a supportive environment for the training of African Americans in higher education, especially in science and engineering. Established in a segregated society, these colleges continued to suffer oppression and lack of adequate funding throughout the twentieth century. As a result of having to defend themselves against such careful scrutiny and deal with concomitant poor funding and limited resources, HBCUs are continuously reinventing themselves to stay afloat. While they strive to remain true to their mission, they are sometimes limited in the range of programs and services they can offer their students. Nevertheless, until recently HBCUs produced most of the expanding pool of graduates in science, engineering, and health-related fields.

These institutions have undoubtedly made a difference in the scientific education of African Americans. Their unquestionable success in providing mentoring and nurturing in addition to academic services is attested to in the literature. According to Trent and Hill's study (1994), HBCUs produce about 30 percent of the African American baccalaureate graduates in science and engineering and about 15 percent of African American master's degree graduates in the physical, biological, and agricultural sciences. This feat is outstanding when one considers that HBCUs represent only 4 percent of the master's degree–granting colleges and universities. Furthermore, they provide a sizable percentage of graduates who enter the pipeline for doctorates in science and engineering. About 43 percent of the African Americans who graduated with doctorates received one or more previous degrees from HBCUs. If it were not for HBCUs, most of the early strides in increasing the number of African Americans receiving degrees in science and engineering would simply not have happened. HBCUs are clearly producing a sizable percentage and number of science graduates, some of whom enter the pipeline for advanced science degrees. Although under constant criticism and denigration, the role of historical black colleges in educating African American women and men of the past and future cannot be dismissed.

If we examine current data just on the number of bachelor's degrees received by black women and black men, we see that some fields still have more black men graduates than women (tables 1–3). Engineering and sometimes the agriculture sciences are still dominated by black men, even at HBCUs. Significant gains have been made by black women at the bachelor's level in the biological sciences and computer sciences at HBCUs (table 2), but the number of black women receiving doctorates in engineering has remained low (table 4). Two HBCUs, Spelman College and Tuskegee University, have been leaders in promoting science and engineering for black women. Like Tuskegee, Spelman College was founded in 1881, but its mission was and remains the education of black women. Even so, Spelman did not produce vast numbers of black women in science initially. During the early 1970s, Dr. Albert Manley, the late president of Spelman College (1953–1976) started a series of summer programs to improve opportunities for and to encourage African American women in the natural sciences. As Manley describes in his memoir, *A Legacy Continues . . .*, he launched the summer programs in 1973, and 50 women students from across the U.S. enrolled. The summer program helped to change the number of science majors at Spelman, from less than 10 percent of the student body in the 1960s to more than 50 percent by the time of Manley's retirement in 1976. In a study by Leggon and Pearson (1997), Spelman and Bennett Colleges were cited as two historically black women's colleges active in producing scientists with doctorates. Spelman College was regarded as one of the most productive undergraduate institutions, producing black female doctorates over a 27-year period. Dr. James Henderson, professor emeritus of biology and emeritus director of the Carver Research Foundation at Tuskegee University, describes a similar impetus launched at Tuskegee (personal communication, 1997). Henderson recalls having the first African American female biology major in the late 1950s and the science faculty's continuing efforts to attract more females to the science program through summer enrichment as well as research opportunities provided by the Carver Foundation. Certainly other HBCUs have similar stories to tell, but Spelman and Tuskegee represent two early efforts. The role that HBCUs have played in producing African American women scientists and engineers is both exemplary and visionary.

Early Influences in the Life of Black Women Scientists

What happens when historians leave out many of America's peoples? When someone with the authority of a teacher describes our society, and you are not in it? Such an experience can be disorienting—a moment of psychic disequilibrium, as if you looked into the mirror and saw nothing.—Ronald Takaki, *A Different Mirror: A History of Multicultural America*

Most young black girls looked into history books and never saw themselves. With the exception of George Washington Carver, most of them never heard their teachers speak of successful black scientists. If they were fortunate, they may have heard of Madame Curie, but most never heard of any female scientist. It is as if the world of science as a possible career choice did not exist for them.

As a young black girl growing up in rural Alabama in the 1960s, I never imagined myself being a scientist, but I did know I would do something special with my life. My three s-heroes before the age of six were my great-grandmother, grandmother, and mother. They did everything on the farm where I lived, from plowing the fields to balancing accounts and keeping accurate farm records. As a result, they showed me by example and through nurturing my dreams that I could do anything that I wanted to do, traditional or nontraditional. It was only later in my educational career that I learned that being a soil microbiologist was considered a nontraditional, male-dominated field and no place for a black woman like me. Fortunately, by that time anyone else's opinion about my career choice didn't matter, because the seed of who I could become had already been planted.

Black girls do not choose science as a career for several reasons. Many young black girls and women still do not have a clear sense of possible career choices, or they fear science and math (especially once they reach high school). Some get distracted, as they have not been taught how to appropriately channel their developing energies and interest in the opposite sex, and still others lack an image of themselves as scientists and engineers, along with a host of other reasons. How, then, did the few who entered science manage to overcome or ignore the "psychic disequilibrium" of which Ronald Takaki so aptly spoke? The women scientists I interviewed did what most black women achievers have done. They looked within themselves and saw what nurturing parents, teachers, and a loving community wanted them to see and think about themselves: that with diligence and strength they could be anything they dreamed. Through the eyes of a supportive family and community, these young black girls learned to see beyond what was missing in the textbooks and in the general educational curriculum. They learned to see with their hearts and develop the courage to succeed.

Like me, the black women featured in this volume unequivocally affirmed that a supportive family or teacher was key to their success and interest in science and engineering. Beyond their immediate family, the black community at large played a pivotal role in encouraging and supporting them in their aspirations towards a scientific career. Elvira Doman describes her Sunday School teacher as being influential in teaching her and her peers about the contributions of blacks in American society during the 1940s. This gave them a sense of pride and the belief that they could be successful at any goal that they might hope to achieve. In addition, most of the women's aspirations to succeed were reinforced by dedicated and committed teachers. Their teachers saw it as their mission in life, and many of these scientists believe that is what is lacking in the educational system for young African Americans: commitment and high expectations from both parents and teachers. Evelyn Boyd Granville, a mathematician, best describes this sentiment in her interview when she states: "I would never 'sass' my mother the way I see children do these days. Parents would support you but they expected and demanded respect." Similarly, she states, "My teachers had high expectations." In her experience, a child's race or lack of privilege was deemphasized, and one's ability to succeed in spite of segregation was the most important focus. These sentiments were echoed by all of the women interviewed. As a result, they rose to the occasion to achieve and be successful in their chosen scientific disciplines.

Another important set of factors which played a role in the degree to which black women participated in science careers are class, place, and time of birth. Granville, Dolores Cooper Shockley, Rubye Torrey, and Yvonne Young Clark grew up around the same time (the 1930s and the 1940s). Shockley, Torrey, and Clark clearly came from privileged backgrounds. Both Torrey's and Clark's fathers were professionals (a mathematician and a physician, respectively) in the black community. Shockley, who also came from a middle-class background and grew up in Mississippi, recounts that she wanted to become a pharmacist at an early age, so her mother had the foresight to send her to a private missionary school, where she could take the appropriate preparatory courses. On the other hand, Granville, who came from a more working-class background, reveals that growing up in Washington, D.C., where she had access to libraries (even though it was during segregated times), caring parents, and teachers with high expectations, made a big difference. She recalls that she was expected to go to an Ivy League college after high school. Doman, who grew up a few years later, also recounts that being raised in a city with access to public libraries probably made a difference in her ability to choose a scientific career.

Regardless of class, time of birth, or geographical location, most of the interviewees cited having a natural curiosity as a reason for their selection of science. In most instances, curiosity when nurtured by supportive parents, and teachers helped to enhance their interest in science. For example, some of the women talked about their parents' purchasing chemistry sets or helping them design science fair projects, which encouraged them early on. As they matured, teachers played a major role in rewarding and encouraging them to pursue science as a career. This combination of factors influenced these women early on to pursue a scientific career.

The Role of the Civil Rights Movement and the Women's Movement

It is important to remember that the rise in the number of black females entering science and engineering parallels the rise of the civil rights movement, the women's movement, and affirmative action programs. The 1960s were years of

Black woman scientist being mentored in an agricultural science lab around 1945 at Tuskegee University.

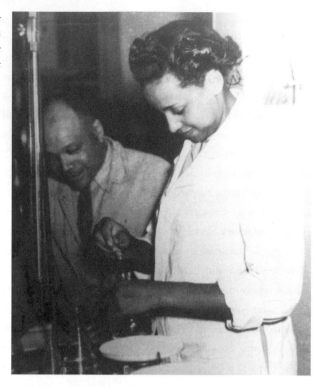

great social change for the nation, with the Vietnam War, marches on Washington and sit-ins for equal rights and access, and the assassination of Dr. Martin Luther King, Jr. and the Kennedy brothers. The nation was also in turmoil on a number of social issues. President Lyndon Johnson sought to redress educational and social inequities by supporting legislations that directly affected the higher education of black citizens. Along with the passage of the 1964 Civil Rights Act, the 1965 Higher Education Act provided the first general federal undergraduate scholarships in the history of the nation (Clewell and Anderson, 1995). The Higher Education Act provided for work-study programs and the TRIO program, including Talent Search and Upward Bound, which targeted minority and disadvantaged students. For African American students like Lynda Jordan, featured in this volume, a program like Upward Bound helped to save her from the streets of Boston. Jordan recounts that she was "hanging out and maybe looking for a little trouble in the streets" before she met her mentor at the Upward Bound Program.

According to Clewell and Anderson (1995), between 1954 and 1969, enrollment at private black colleges increased by about 90 percent, from 25,569 to 48,541. This trend in the increased enrollment of African Americans was seen in both HBCUs and PWCUs. Other affirmative action programs also helped to boost the enrollment of African Americans, women, and other minorities in colleges and universities. Interviewees born in the late 1940s and the 1950s and thereafter strongly believe that the civil rights movement played a vital role in providing opportunities for success in scientific fields.

Most of the women born before this time believe that their success had been realized before the 1960s. They are essentially the pioneers in their specific disciplines who paved the way for women and minorities after them. Some of these women were the first African American females to receive doctorates in their chosen fields (such as Dolores Shockley in pharmacology) as well as at the bachelor's level (such as Yvonne Y. Clark in mechanical engineering).

While most of the interviewees saw the civil rights movement as significant—even if it did not directly affect their advancement in science—most black women do not identify with the women's movement, because they see it as a white-woman-only movement that failed to address the concerns of black women. Not surprisingly, the black women scientists who were asked

Geneticist Dr. John Williams and a student at Tuskegee University, 1972.

about the women's movement said that it had not made an appreciable difference in their professional lives. (A few of the early interviewees were not asked
this question because it was added later.) Only a small number of the women
interviewed felt that the women's movement had mattered in their personal
and professional lives. We both share a history of pain and suffering, but African American women have the burdens of both race and gender and often
feel misunderstood by their white counterparts. Alice Childress, one of the
writers in S. M. Jordan's book *Broken Silences: Interviews with Black and White
Women Writers*, captures the sentiment of many black women who still view
the women's movement as a white woman's movement:

> If you would simply say that we will be a full and equal part of
> the women's movement, I think that would attract a lot of black

*Dr. Lynda Jordan works in the biochemistry laboratory with her graduate stu-
dent, Stephen Webb, at North Carolina A & T University.*

women. I don't belong to NOW (National Organization for Women). I don't object to it. It's a good idea but they don't seem to have enough interest in black issues. There are some issues that are particularly black and they feel everything else that is not particularly black is more important. They figure we can all join hands on things that are not particularly black; it's for anybody. But we are all particularly somebody with very particular needs.

For Lynda Jordan and Hattie Carwell, the women's movement does not take into account the special concerns of black women. As with other professions, some white female scientists may simply not view their oppression as different from black female scientists'. They may not "get" her problem, since by today's standards a black woman is seen as a bonus hire due to her race and gender (see the race and gender discussion, next section). Many black women see the women's movement as setting the agenda and, as Alice Childress describes, expecting black women to follow what has already been planned without giving any consideration to differences. This attitude, whether real or perceived, alienates some black women, including many black women scientists. Other studies and articles on higher education support this view of how many black women feel about the women's movement (Moses, 1989; Bernstein and Cock, 1994; Jordan and Ford-Logan, 1994; Brown, 1995). In Bernstein and Cock's article in the *Chronicle of Higher Education*, "A Troubling Picture of Gender Equity," they discuss how the issues of minority and underprivileged women have been subjugated to the experiences of white women. According to these authors, the white woman's experience has become "universal" without regard to race, ethnicity, socioeconomic class, and age group. This approach of "lumping" black women or other minority women in the women's category or minority category creates a distorted picture of what is actually true about the black woman's status in higher education, careers, and society. This distortion may often lead to inadequate counseling and resources, but more often to inaccurate policy- and decision-making.

Nonetheless, a few of the black women interviewed believe that they have directly benefited in their careers as a result of the women's movement. Clearly, there is still a lot of work and bridge-building that needs to occur between black women and white women, including those in the scientific professions, where there are fewer women to build coalitions. I don't pretend to speak for all black women, but in my research and personal conversations with hundreds of black women over the last 20 years, some clear feelings emerge. Briefly, many black women simply don't feel welcome in the women's movement. Many black women view it as a movement for privileged white women who ignore their concerns. Whether these feelings are justified or unjustified, they divide all women on common issues and keep them from working toward what should be collective efforts for both groups.

Race and Gender Issues for African American Women Scientists

> The colored woman of today occupies, one may say, a unique position in this country. In a period of itself transitional and unsettled, her status seems one of the least ascertainable and definitive of all the forces which make for our civilization. She is confronted by a woman question and a race problem, and is as yet an unknown or unacknowledged factor in both.

These words, written by Anna Julia Cooper in 1892, describe the position of black women in American society. Over 100 years later, black women are still faced with the issues of her race and gender in the home, in the workplace, and in society. Equally important but often not discussed is how class affects the lives of African American women. When asked if race influenced their career choice or success in science, most black women said "yes." Most of the scientists believed that race had been a significant factor in how they are viewed, not just as a scientist but also as people in American society. The responses varied, however, from stories of overt racism to stories in which race actually worked in their favor. As might be expected, the women who were born before World War II were more likely to have experienced racial barriers as they tried to enter their professions or receive graduate training. Most of the women who were being trained during the 1940s and the 1950s could easily relate a story about how they were denied their rights due to a segregated society. Dolores Shockley even describes the difficulty of finding off-campus housing in West Lafayette, Indiana, in the 1950s. Elvira Doman and Geraldine Twitty (of the same generation) and Georgia Dunston tell stories of being denied jobs because of their race early on in their careers. These women sought employment as laboratory technicians in the Northeast but were told the only jobs available were for dishwashers or maids. In several cases, this racially inspired denial of employment caused the women to seek further education.

However, there is an interesting twist in which race in fact may have favored the black woman scientist during segregation. Evelyn Boyd Granville, who was in graduate school during the 1940s, talks about how she was viewed as an anomaly during her time at Yale and thereafter. She recalls, "There were five or six of us (women) working on doctorates, but only two of us received the Ph.D. The rest (white women) received master's degrees." Shockley also relates a similar story of some faculty members who did not expect the white women to finish their doctorates at Purdue University in the 1950s. These women were expected to get married and not actually practice their science. She further recalls in her own case, "I guess they didn't care about what I did. I am not really sure." This is an excellent example of a case where race certainly played a role in leading to the scientist's successful completion of a doctoral program. Perhaps because the faculty had no concrete expectations of this black woman and she appeared to be of no particular threat to the establishment, no insurmountable barriers were placed in her way.

The expectation that women would not become practicing scientists was not limited to the white, male-dominated research institutions. Twitty recalls similar sentiments at Howard University during her undergraduate days. She states, "The attitude was that women in the sciences were looking for successful husbands."

Because some gains had been made in the 1960s and the 1970s due to the civil rights movement and the women's movement, the responses are more varied for the younger generation of scientists as to how race played a role in their lives. For example, Etheleen McGinnis-Hill was reared in Birmingham, Alabama, during the height of the civil rights movement. She relates very positive experiences of her graduate work at Purdue University in the 1970s, including the fact that she had a black graduate advisor during her stay. She did not encounter the same difficulties in finding housing as Shockley had experienced nearly a generation earlier. On the other hand, when Jennie Patrick arrived at Berkeley as an engineering student, she contended with both race and gender issues.

When asked if they could separate out the race and gender issues, the women's responses become even more ambivalent. Although some black women scientists believe that race is a primary factor in how they are viewed, the discussion during the interview and afterwards did not always bear this out. La-Vern Whisenton-Davidson puts it best when she says that "it depends on who is doing the viewing as to whether or not I think it's my race or gender at question." Some of the black women work in predominantly white institutions and some work in historically black institutions. To some extent, each of these workplaces colors the experiences of racism and sexism. Generally, black women scientists at historically black institutions did not see race as a major issue, and many cited gender issues as a real problem. An exception was Anna Coble, a physicist, who saw race as being more of an issue early on; although she was at a historically black institution her department was primarily composed of white male faculty. This is an important point about the composition of science and engineering faculty in historically black colleges and universities. Many people falsely believe or assume that white males do not hold key supervisory or authority positions in predominantly black institutions and think that race shouldn't be a major issue in these settings. That is not always the case. In fact, most black institutions in the early part of their history were run by white men (as presidents and trustees) and some white women. Many HBCUs still use the earlier models of leadership. Spelman College, for instance, did not get its first black female president until 1987, when Johnetta Cole was selected after an outcry from its all female student population—over 100 years after Spelman College had been founded. In other words, the residuals of those early models, for good or bad, often continued to exist a hundred years after their inception.

Gender issues within race is a topic that many black women (including black women scientists) are reluctant to comment on publicly, especially as it relates to problems within our own race. Dolores Shockley, Lynda Jordan, and Georgia Dunston's comments were an important contribution to opening this overdue conversation. They articulated the unequal treatment of black women at some historically black college campuses and the effect that such treatment can have on producing and retaining young and older black women in science.

Black women have always spoken out, but have their voices been heard? According to Deborah Gray White's *Too Heavy a Load: Black Women in Defense of Themselves 1894–1994*, black women have long spoken out on the issues of gender discrimination within the race. She recounts the story of Amy Jacques Garvey, the second wife of Marcus Garvey, in her response in the 1926 Black Women's Resolve. Jacques Garvey wrote: "We serve notice on our men that Negro women will demand equal opportunity to fill any position in the Universal Negro Improvement Association or anywhere else without discrimination because of sex." Garvey clearly conveys the need for their rights as women to be heard among black men.

In recent times, Professor Anita Hill and Justice Clarence Thomas brought forth again the real dilemma and problems the black woman faces when she speaks out about gender issues within the race. Cole and Sheftall (2003) in *Gender Talk* place the issue in context. I paraphrase some of their discussion:

The issue of race loyalty becomes a prominent feature in public discourse. There were profound differences between black women and men and among black women about how to deal with these public crises such as the Thomas-Hill Supreme Court hearings, the O.J. Simpson trial, the Mike Tyson saga, and others. These highly publicized cases forced the black community to respond to what they perceived to be harmful airing of dirty linen in public.

The ideology of race loyalty first and gender second has been passed down for generations. Because the black woman knows there are real consequences for anything negative that she might publicly say about black institutions or black men, she is often silent or less vocal on gender issues within the race.Lynda Jordan speaks openly about her concerns for speaking candidly about historically black colleges and universities. However, the ultimate cost of not fully addressing these issues and bringing the conversation about the needs of black women full circle effectively silences the black woman scientist or severely limits her ability to address concerns in a way that could benefit herself, the scientific community, and the nation as a whole.

In predominantly white institutions, race may be seen as more of an issue for black women scientists, but the picture is still not clear when one attempts to clarify the differences. LaVern Whisenton-Davidson, who works at a mid-sized, predominantly white state institution, attempts to clarify these issues of both race and gender:

> White colleagues, both male and female, don't realize that you are less accepted because of your race and gender. Because the white male is considered to be superior in American society, he is automatically accepted. While it is true that white women must deal with gender issues, she is not confronted with the issues of race. That is also true for the black man. He has one issue to deal with in the larger context of society.

In the academy, black women must deal with both these issues throughout their educational and professional careers. E. M. Ellis (2001), in a study on the impact of race and gender in doctoral programs, found that black women were generally more dissatisfied with their experiences, partly due to their race and gender. Black women tend to feel more alienated and dismissed in classroom settings on predominantly white campuses than white men or black men, in part because males seemed to form alliances with other males in their departments. Some attempts at alliances with white females worked well but still did not yield a positive relationship where the black female felt equally valued, because she felt that the white female did not want to deal with the issues of race in the everyday lives of a black woman. These sentiments are echoed by many black women professionals both inside and outside of the academy.

The issues of the black women scientists tend to fall through the cracks because of her dual identity. Some white women do not recognize that they are privileged because of their race. This is not to diminish their experiences of gender discrimination in science, but the white woman is still seen as an asset because of race in American society.

On the other hand, black men are also privileged, because of their gender. Often I find in my conversations with black male scientists that they like to lump all the issues of black women under race. Some of them are hard pressed to separate the issues of race and gender. There is no question that black men have suffered inhumane treatment in America; I don't think any of these black women that I interviewed would disagree with that statement. We women often have discussions of the emasculation of black men in society as well as how our young black men are steadily filling the jail cells of America. Notwithstanding all these pertinent issues in our black community, black men would be wise to listen to and be open to his sisters' stories. Until they, as well as other minority groups, fully recognize the dual position of black women, the contributions and retention of black women in the professions will never be fully achieved. These "isms," as well as other differences (disabilities, sexual orientation, etc.) will probably continue to cripple a whole generation of untapped scientific talent.

Hattie Carwell, who works for a government agency, describes race as a factor which influenced her promotion and benefits in her career. She further states that gender became more of a problem as she progressed through her science career to a senior position. Everyone wants those senior positions, suggests Carwell, but unless you have been groomed like the males, your possibilities for success are limited. In other words, the glass ceiling became more of an issue when Carwell rose to a position in which she could influence policy decisions in the organization.

Still, there are black women scientists who do not view race or gender as a problem in their careers. In fact, they view both race and gender as positive factors in their lives. Evelyn Granville's interview brings this issue to light when she says, "I never viewed it as a problem or thought about it. I let it be the other person's problem." In my earlier publications on black women in the agricultural sciences, I found similar responses among a few of the women, especially the younger scientists. I admittedly pressed Granville on the question because I knew she grew up during a time when segregation and discrimination were rampant in the United States. Although she stood her ground, she later stated that her pioneering status and the time in which she grew up may have affected her view, and how others viewed her as well. My own motives for posing further questions to Granville was not in any way to diminish her positive experiences but more to shed light on how a black woman mathematician had been so successful in navigating a science career through obvious times of segregation. She was a young black woman armed with high self-esteem, excellent skills in math and science, and a determination to succeed. Simple ingredients, but hard recipes, to impart to today's youth. How to sort out which factor, race or gender, is truly affecting one's career and life is and has been an issue for black women for over a century. There are both positive and negatives experiences that can be recounted by these and other black women scientists. As I try to sort out some of the differences in this book, I can only hope that in 2096 young black women will not have the same issues or questions.

Balancing Family and Career

Another factor in the success of the black woman scientist is finding the proper balance between maintaining family and a career. Most of the interviewees stated that a supportive spouse or network of family and friends was important in a science career. These women urged women of any race to choose a spouse very carefully, especially if they are doing so early in their career. Stories were related both here and in private about how some careers have ended early for young black women because of poor selection of mates. Granville recounts that she did not make the best choice in her first husband, but she did not repeat that mistake. She found a loving and supportive mate in her second husband.

When discussing the impact of childbearing on tenure and promotion, the women can be separated into two groups. Three of the women did not experience this issue because they are single and childless. The first group of women, those who were born before World War II, seemed to have dealt more traditionally with childrearing and family roles. Shockley, a pharmacologist, talks about spending most of her paycheck on babysitting. Geraldine Twitty, a zoologist at Howard University, tells how she would get to the lab by 5:30 to get her research going before she had to teach her 8 A.M. class. She stated, "It was hard, but where there was a will there was a way." Women who came of age in the 1960s and 1970s seem to have had a few more options, depending on their particular situation. For example, Freddie Dixon's husband clearly provided much of the care for their young son during her graduate studies. Still, some of the women, like LaVern Whisenton-Davidson and Anna Coble, chose to marry later and found very supportive and caring spouses. Both married and unmarried black women scientists echoed the same sentiment: young black women need to choose supportive mates or significant others carefully.

The Culture of Society and Science: Demystifying Stereotypes and Marginalization of Black Women Scientists

Image is everything. How does a scientist look? Ask any elementary class this question and see what answer you will typically get. For that matter, ask any freshman college class and see what the typical answer will be. Typically, the response is still that it is a white male with glasses wearing a white coat. They see Einstein or any number of cinematic portraits of scientists dressed in typical scientific garb. Most women are not perceived to be scientists or engineers, but for black women the situation is even worse. How does this stereotype play itself out in the everyday life of the black woman scientist? I would like to share a story to illustrate this point.

In the summer of 1996, I flew to Washington, D.C., to work on a research panel for a federal funding agency. We, as a panel, would decide whose research would get funded. The plane was full and I was sitting in the aisle seat, and a white gentleman sat next to me. As we were comfortably flying at about 30,000 feet, I decided to take one final review of the work that I had completed on the proposals. As I read, I became aware that the gentleman seated next to me was reading along as well, and I gave him a somewhat cold stare. He replied, "Oh, I am so sorry, but I like to see what people are reading and I am fascinated by what you are reading." Of course, I had been down this road so many times that I knew he was more fascinated by *who* was reading than by *what* was being read. I smiled because I knew what his next words would be. We talked for a few minutes about what I was reading, and as predicted, he uttered the words that had become so familiar to me. He said, "I hope you will not think that I am sexist or prejudiced, but you don't look like a scientist." My first inclination was to be sarcastic and reply, "Well, what does a scientist look like?" However, I just couldn't pass up another great teaching moment. Besides, he seemed like a decent human being whom I would probably never see again. To make a long story short, we had a nice conversation about my profession and his as well. I have experienced many versions of this conversation over the last 25 years. The image and perception of who a scientist is is not someone who looks like me. Many women have described similar scenarios, both in this volume and in my research.

If black women scientists are to become partners in the scientific and engineering professions, the image and perception of who a scientist is must be changed in this society. There have been some efforts to make minorities and women in general more visible. As usual, black women have fallen through the cracks. While the black woman must share in the responsibility for her own visibility, more must be done to bring the diversity and creative ability of black women to the forefront. Where possible, we must showcase and even lobby Hollywood for more diverse images of black women in the movies. Our society is media-driven, and that is one way to give young people a glimpse into the occupations of black women, including scientists and engineers.

"Isolation within Isolation": The Marginalization of Black Women Scientists

One of the key factors affecting black women who are in the pipeline for science and engineering careers is dealing with isolation. Read any study on race and gender issues in higher education or corporate America, and isolation is usually cited as one of the main issues for black women. The black woman scientist has a particular problem: because she generally doesn't have a support system in place, she has to create one for herself *and* prove that she is worthy of the training and career in science. Evelyn Boyd Granville's experiences at Smith College in the 1940s were unusual then, and still today in many scientific circles. Shirley Jackson, who was a brilliant young physics student at MIT during the 1960s, notes the initial isolation she experienced as an undergraduate—until it became clear that she was at the head of the class.

Collaboration is an essential part of the culture of science. Despite the media portrait of the wacky, off-the-beaten-path scientist who is in some lab by himself creating some strange concoction in a test tube, this is not the way that most scientific training and investigation work. The difficulty of developing collaborations in scientific research and forming social networks are factors that discourage women and, quite frankly, some young men from completing an advanced degree in science.

This isolation was very real to me as the only black woman in my graduate program when I attended an HBCU and in my doctoral studies at a predominantly white institution. I knew the numbers were not on my side as far as race and gender were concerned. I sought out networks and assistance from other sources on- and off-campus. Many of the women interviewed for this book expressed the need for such an approach to succeed in science. The social and cultural isolation experienced by some black women is often the breaking point of a scientific career, even today. In the face of this isolation, many young black women view a career in science—which often does not pay off financially or emotionally—as not worth the trouble. In some ways, black women who matriculated during the Jim Crow era found that it was easier to deal with the isolation, provided they had been academically prepared. Granville's, Shockley's, and Clark's stories bear this out. The lines of segregation were clear-cut and the goals of education were very clear in these women's minds. They were representatives for the black race, and they knew how to cope. I believe they were actually pursuing a goal much larger than themselves. Women who studied during or after the civil rights movement sometimes tell a different story of their experiences. Lynda Jordan talks candidly about feeling as if she had to prove herself over and over again. Legal segregation has ended for today's young black woman, but the issues of navigating a career through the double burden of race and gender remain. She has to create a support system for herself in order to achieve success in a scientific career. White men, in particular, have a built-in support system, the "good ol' boys" network. Some black women scientists must deal with the isolation from "mainstream" scientists and engineers as well as some distance from scientists of her own race (black men, for any number of reasons) and of her own gender (white women and other minority women). This leads some black women scientists to experience what I coin as an "isolation within isolation" syndrome.

Because of the black woman's history in American society, other groups subscribe to many stereotypes of the black woman, whether consciously or not, and this scenario is played out in the workplace. Some of the common stereotypes of the black woman—as a "mammy figure," the maid, or "Sapphire"— are indelibly etched in the minds of other Americans and are too often the lenses through which she is viewed and treated. When the black woman scientist refuses to be placed in one of these stereotypical roles, she often has hell to pay for not conforming to these images. The result may be a lack of cooperation on team projects and/or lack of recommendations or recognition for her work. In the academy, where black women can have a significant impact on scientific research, they are often overloaded with a disproportionate amount of university and community service, which affects their ability to progress through the academic ranks to a successful science career. These are crucial issues when the black woman scientist has to work in a collaborative mode with other scientists, who are still mostly white and male. In other professions, such as education and some social science disciplines, the number of black women professionals is greater and continues to grow, so there is more opportunity and likelihood that she will find like minds to work with. This is not to suggest that black women scientists don't find healthy, productive relationships with other scientists. They in fact do, with both men and women of all races. However, it can be a real issue for black women scientists when success in science is so dependent upon collaborative research. Too often she has to wait patiently for her counterparts to realize she is not a stereotype but an individual with similar needs and aspirations.

Black women who work in predominantly minority or women's institutions sometimes have similar problems, but for different reasons. On the black college campus, some black male scientists view their sister as a threat and often resort to unhealthy competition rather than cooperation with her. I have been told about many of these incidents "off the record." Of course, most black women feel that they have otherwise been embraced by the black experience. Thus this insecurity, whether founded or unfounded, creates stress for black women in science.

Another thorny issue that the black woman scientist must deal with is her acceptance as a scientist within her own culture. Most black women know how to build a social network outside the university and workplace, provided it is available in her environment. Lynda Jordan is remarkably candid about not being understood or accepted by other blacks as a scientist, and this is an issue that we have to discuss. A young black woman scientist (not included in this book) stated to me, " I am not accepted by my own family because I want to be a scientist. They think that it is weird that I want to pursue a career in natural sciences." This young woman is on her way to becoming a scientist, but her family's lack of understanding makes it difficult for her to cope. Anna Coble touches on this problem when she points out that many young black women explain to her that their boyfriends do not want them to study physics. She warns them to be careful in the choices they make, as they will have long-term implications in their success in science. While other women may experience similar issues, black women and native American women are the least likely to choose and continue in a career in science. Therefore, a discussion of these concerns is important in addressing her success in science.

My grandmother Marie, who was born in 1914, had more opportunities than her mother, Nettie Jay, in 1894. She completed the eighth grade, the highest grade level in her town. She spent her entire life farming, without a day of formal training. Yet I dare say that my grandmother, who loved and nurtured the earth, knew it far better than I, who have earned a Ph.D. and a postdoc in soil microbiology. It is on the shoulders of these great women that I stand. I tell some of the stories of women in my profession who have endured and contributed substantially to the research, economy, and culture of our great country.

Black women bring a unique blend of culture, strength, courage, character, and outstanding skills and analytical abilities to the table. We should be appreciated and welcomed to partake and give in the scientific community as we help to maintain and attract a whole new generation of scientists and engineers in the new millennium. I hope this volume on black women scientists will whet your appetite for more of our history and the science that we do in our daily lives.

TIMELINE

Black Women in Science, Mathematics, and Engineering

Events in the history of science, black history, and women in medicine and health sciences are included to give the reader additional context.

This timeline represents the best information available from books and articles, websites, and personal communications. There may be several black women not included in this list who have accomplished significant achievements in science, engineering, and technology and who may have also been the first to do so in a particular field or scientific discipline. This timeline is an effort to recognize some well-known and not so well-known scientists, firsts in medicine, dentistry, and pharmacy, and some of the historical events surrounding these achievements. Any comments or suggested references may be sent to the publisher for consideration in future editions of this book.

Date	Event	Place
1619	Twenty Africans, three of them women, are put ashore off a Dutch frigate.	Jamestown, VA
1624	Massachusetts is the first colony in North America to give statutory recognition to slavery.	Boston, MA
1787	The U.S. Constitution, with three clauses protecting slavery, is approved at the Philadelphia Convention.	
1793	The female Benevolent Society of St. Thomas is founded by black women.	Philadelphia, PA
1824	Renesselaer Institute is established to train persons in science and its applications "to the common purposes of life."	Troy, NY
1835	Oberlin College becomes the first U.S. college to admit students without regard to race or gender.	Oberlin, OH

1840s	Women begin creating their own scientific societies, such as the Female Botanical Society of Wilmington. It is not known if black women were a part of this group—presumably they were not.	Wilmington, DE
1861	The Civil War begins.	
1862	The United States Department of Agriculture is formed.	Washington, DC
	Mary Jane Patterson earns a B.A. degree from Oberlin College, making her the first black woman to earn a bachelor's degree from an accredited U.S. college.	Oberlin, OH
	The First Morrill Act, supporting white land-grant colleges and universities, is passed.	
1863	The Emancipation Proclamation frees the slaves in those states rebelling against the Union.	
1864	Rebecca Lee (Crumpler) becomes the first black woman to graduate from a U.S. college with a formal medical degree, from the New England Female Medical College.	Boston, MA
1865	The Thirteenth Amendment to the U.S. Constitution, abolishing slavery, is adopted.	
1867	Howard University is founded.	Washington, DC
	Rebecca Cole becomes the second black woman to receive a medical degree in the U.S., from Women's Medical College of Pennsylvania.	Philadelphia, PA
1873	Bennett College for black women is founded.	Greensboro, NC
	Susan Smith McKinney Steward becomes the first black woman doctor to be formally certified in New York and is later founder of the Women's Loyal Union of New York.	New York
1876	Meharry Medical College is founded as the Central Tennessee College to educate blacks in medicine.	Nashville, TN
1879	Josephine A. Silone receives a degree from Rhode Island Normal School, presumably a science degree.	Providence, RI

	Mary Elizabeth Mahoney becomes the first black woman to graduate with a nursing degree, from New England Hospital for Women and Children.	Boston, MA
1881	Spelman College for black women is established.	Atlanta, GA
	Tuskegee Institute (University) is established.	Tuskegee, AL
1884	Anna Julia Cooper graduates from Oberlin College. (Cooper later becomes a leading advocate and educator on race and gender issues. She is probably most known for her book, *A Voice from the South*.)	Oberlin, OH
1885	Sarah Goode is the first black woman to receive a U.S. patent, on July 14 for her folding bed.	
1888	Miriam E. Benjamin receives a patent for her gong and signal chair.	
	Josephine Silone Yates becomes the first black woman to head a natural science department, at Lincoln University.	Jefferson City, MO
1890	Ida Gray is the first black woman to graduate with the doctor of dental surgery degree.	Ann Arbor, MI
	The Second Morrill Act is passed, to help establish land-grant colleges for blacks.	
1893	Georgia Patton is one of the first black women to earn a medical degree from Meharry Medical College.	Nashville, TN
1895	The National Medical Association is founded.	
1896	The Supreme Court's decision in *Plessy v. Ferguson* affirms the concept of separate but equal.	
1903	Beebe Lynk receives a degree, presumably in pharmaceutical chemistry, from West Tennessee University.	Memphis, TN
1904	Mary McCleod Bethune establishes what later becomes Bethune-Cookman College.	Daytona Beach, FL
1909	The National Association for the Advancement of Colored People (NAACP) is founded.	

1921	The first black women earn Ph.D. degrees throughout the U.S. (none were noted in science or engineering).	
1922	Bessie Coleman, the first black woman aviator, gives her exhibition on Long Island.	Long Island, NY
1926	The National Technical Association is founded by black men to advance careers and educational opportunities for black engineers and scientists.	
1928	Georgia Caldwell Smith receives a B.S. degree in mathematics from the University of Kansas; in 1929 she earns a master's degree.	Lawrence, KS
	Marjorie Stewart Joyner receives a patent for the permanent wave machine.	
1933	Ruth Ella Moore receives the Ph.D. in bacteriology from Ohio State University, likely the first black woman doctorate in science.	Columbus, OH
1935	Jessie Jarue Mark was the first black and woman to earn a Ph.D. in botany from Iowa State College (now University).	Ames, IA
	Flemmie Kittrell was the first black woman to earn a Ph.D. in nutrition from Cornell University.	Ithaca, NY
1939	Tuskegee Institute establishes a school of nurse-midwifery.	Tuskegee, AL
1939	The Manhattan Project is developed. This wartime effort was designed to build the first atomic bombs.	
1940	Roger Arliner Young is the first black woman to earn a Ph.D. degree in zoology from the University of Pennsylvania. (Previously, she was the first black woman to work as a scientist at the Marine Biological Laboratory in Woods Hole, Massachusetts, in the late 1920s.)	Philadelphia, PA
1941	Marie Clark Taylor receives a Ph.D. from Fordham University in botany.	New York, NY

| 1941 | Ruth Smith Lloyd is the first black woman to receive a Ph.D. in anatomy from Case Western Reserve University. | Cleveland, OH |

| 1942 | Marguerite Thomas receives a Ph.D. in geology from Catholic University. | Washington, DC |

| 1943 | Euphemia Lofton Haynes is the first black woman to receive a Ph.D. in mathematics, from Catholic University. | Washington, DC |

| 1944 | Mary Logan Reddick becomes the first female instructor of biology at Morehouse College. (She eventually earns her Ph.D. from Radcliffe in neuroembryology.) | Atlanta, GA |

| 1945 | World War II ends. | |

| 1945 | Marion Antoinette Myles receives a Ph.D. in plant physiology from Iowa State College (later University). (She was probably one of the earliest black women to receive a B.S. degree in botany/plant science, in 1937 from University of Pennsylvania.) | Ames, IA |

| 1946 | Hattie T. Scott is the first black woman to receive a B.S. degree in civil engineering from Howard University. | Washington, DC |
| | The United Negro College Fund (UNCF) is established to help support black colleges and students. | |

| 1947 | The National Medical Association is founded. | |

| 1948 | Marie Maynard Daly is the first black woman to receive a Ph.D. in chemistry from Catholic University. | Washington, DC |

| 1949 | Evelyn Boyd Granville and Marjorie Browne receive the Ph.D. in mathematics, from Yale University and the University of Michigan, respectively. | |

1949	Alfreda Webb graduates in veterinary medicine from Tuskegee Institute (now University), where she was a member of the Institute's first veterinary class.	Tuskegee, AL
	Jane Hinton receives a doctorate of veterinary medicine from the University of Pennsylvania.	Philadelphia, PA
1950	Cecile Hoover Edwards receives a Ph.D. in food chemistry from Iowa State College (now University). She is believed to be the first black woman to do so.	Ames, IA
1952	Yvonne Clark Young becomes the first black woman to receive a B.S. degree in mechanical engineering from Howard University.	Washington, DC.
1953	Discovery of the structure of DNA.	
1954	Hildreth G. Florant becomes the first professional female member of the National Technical Association.	
	Brown v. Board of Education states that separate but equal schools are unconstitutional.	
1955	Dolores Cooper Shockley is the first black woman to receive a Ph.D. in pharmacology from Purdue University.	West Lafayette, IN
	Rosa Parks refuses to give up her seat on a Montgomery bus to a white man, sparking the modern-day civil rights movement.	Montgomery, AL
1957	The Southern Christian Leadership Conference (SCLC) is formed, with Dr. Martin Luther King, Jr. as the president.	Montgomery, AL
	Dorothy Lavinia Brown becomes the first black woman surgeon in the South. (She later becomes a fellow of the American College of Surgeons.)	Nashville, TN
	The Sputnik era begins.	

* = Information from personal communication.

1960	Lelia Mae Thompson-Flagg becomes the first black woman to receive a B.S. degree in civil engineering from the University of Missouri-Rolla. (She receives her master's in engineering from the University of California, Berkeley in 1962.)	Rolla, MO
	Maxine R. Rosborough becomes the first female officer in the National Technical Association.	
1961	Vivienne Malone Mayes becomes the first black faculty member in mathematics at Baylor University in Texas.	Waco, TX
1964	The Civil Rights Act of 1964 prohibits discrimination in public accommodations and in employment.	
1967	Jane Cooke Wright is the first professor of surgery and first associate dean at New York Medical College.	New York, NY
1968	Rubye Torrey becomes the first black woman to receive a Ph.D. in chemistry from Syracuse University.	Syracuse, NY
1972	Willie Hobbs Moore receives a Ph.D. in physics from the University of Michigan, making her the first black woman to receive that degree in the U.S.	Ann Arbor, MI
1973	Shirley A. Jackson receives a Ph.D. in theoretical physics from the Massachusetts Institute of Technology (MIT), making her the first black woman to do so.	Cambridge, MA
	Anna Coble is the first black woman to receive a Ph.D. in biophysics from University of Illinois.	Urbana, IL
1974	*Shirley M. Malcom receives a Ph.D. in ecology from Pennsylvania State University, making her the first black woman to do so.	University Park, PA

* = Information from personal communication.

1974   Margaret Tolbert receives her doctorate from   Providence, RI
        Brown University in chemistry. (She later be-
        comes known for her research on the biochem-
        istry of the liver and is probably one of the first
        women to serve as director of the Carver Re-
        search Foundation and associate provost of
        Tuskegee Institute (now University).)

1975   Alexa Canady is the first black woman to become
        a neurosurgeon. She received her degree from
        University of Michigan.

        Jeanne Sinkford becomes the first woman dean   Washington, DC
        of a dental school in the United States at Howard
        University.

1976   Increased funding for black land-grant colleges
        and universities is passed (89-106 and Smith-
        Lev.), which might have assisted the land-grant
        colleges in attracting more black females into the
        agricultural science programs.

1977   Hattie Carwell publishes *Blacks in Science: From
        Astrophysicist to Zoologist* with Exposition Press.

        First records of a black woman as a member of   Madison, WI
        the American Society of Agronomy (from Bra-
        zil).*

1979   Jennie Patrick is the first black woman to receive   Cambridge, MA
        a Ph.D. in chemical engineering from MIT.

1980   Shirley Mathis McBay, a mathematician, becomes   Cambridge, MA
        dean of students at MIT.

1981   Jewel Plummer Cobb, a prominent cell biologist,   Fullerton, CA
        becomes president of California State University,
        Fullerton.

1982   *Eunice Fanning Foster receives a Ph.D. in crop   Fayetteville, AR
        science from the University of Arkansas, making
        her the first black woman to do so in that depart-
        ment.

* = Information from personal communication.

1982	*Judith V. Burton receives a B.S. degree in soil science at Alabama A & M University, and later an M.S. degree from Purdue University in soil fertility in 1985. She is the first black woman to do so at Purdue.	Normal, AL
1983	Christine Darden is the first black woman to receive a Ph.D. in mechanical engineering from George Washington University.	Washington, DC
1984	Valerie Thomas becomes the first woman national president of the National Technical Association.	
1985	Edith Irby Jones is the first black woman elected to lead the National Medical Association.	
1986	Lynda Jordan becomes the first black woman to study at the Institut de Pasteur in biochemistry.	Paris, France
1987	Diann Jordan receives a Ph.D. in soil science from Michigan State University. She is the first black woman to do so at that university. (She is later the first black woman to serve on the faculty and receive tenure and promotion in the Soil and Atmospheric Science Department at the University of Missouri-Columbia, in 1996.)	East Lansing, MI
	Mae Jemison becomes the first black woman astronaut.	
1988	Reatha Clark King, a chemist, becomes president of General Mills Foundation.	
	*Dolores Shockley is appointed interim chair of the Department of Pharmacology at Meharry Medical College.	Nashville, TN
	*Georgia Dunston is the only woman to head a National Genomic Laboratory and one of the first black women to head a department of microbiology in the U.S.	Washington, DC
1989	Shirley M. Malcom becomes head of the Directorate for Education and Human Resources Programs, American Association for the Advancement of Science.	Washington, DC

1990	Roselyn Payne Epps becomes president of the American Medical Association.	
1992	Hazel O'Leary is appointed secretary of energy in the Clinton administration.	Washington, DC
1993	Jocelyn Elders is appointed surgeon general in the Clinton administration. She is the first black and woman to serve in this role.	Washington, DC
	*Jamye Carter becomes the second black woman to receive a Ed.D. in mathematics education (Willie Christian was the first, in the 1980s.) Jamylle Carter, Jamye Carter's daughter, receives a Ph.D. in mathematics in 2001 from the University of California. The Carters are one of the few mother-daughter doctoral teams in the U.S.	Auburn, AL
1994	*Dolores Shockley becomes the permanent chairperson in the Department of Pharmacology at Meharry Medical College, making her the first black woman to head a pharmacology department in the U.S.	Nashville, TN
1995	Shirley A. Jackson is appointed chair of the Nuclear Regulatory Commission.	Washington, DC
1997	*Ingrid St. Omer is the first black woman to receive a Ph.D. in engineering from the University of Missouri-Columbia.	Columbia, MO
1997	"Hearts and Minds: Black Women Scientists in the United States, 1900–1960," by Dr. Winnie Warren, is the first dissertation written exclusively on black women scientists, at Indiana University.	Bloomington, IN
1999	*Shuntele Burns becomes the first black woman to earn a Ph.D. at the University of Florida in entomology.	Gainesville, FL
1999	Shirley A. Jackson is selected as president of Renesselaer Polytechnic Institute.	Troy, NY
2000	Condoleezza Rice becomes head of the National Security Council, making her the first black female to ever head such a post in the U.S.	Washington, DC

Ruth Simmons is elected as president of Brown   Providence, RI
University, making her the first black woman to
head an Ivy League university. (She was the former president of Smith College.)

2001   "Sista Scholars Working as Outsiders Within: Tuscaloosa, AL
African American Women Scientists at Majority
Institutions," by Samantha Elliott Briggs, is one
of the first master's theses on black women in science, at the University of Alabama.

2002   *Raynetta Prevo becomes the first black woman   Tuscaloosa, AL
to received a Ph.D. from the University of Ala-
bama in applied mathematics.

2003   Shirley Ann Jackson is selected as president of
the American Association for the Advancement
of Science, making her the first black woman to
head this national scientific organization (term
beginning in 2004).

2004   Advance Leadership Conference, The Status of   Atlanta, GA
Women Scientists at Historically Black Colleges
and Universities, was held in Atlanta, GA, August
6–8. This conference was hosted by Clark Atlanta University and funded through a grant from
the National Science Foundation. To the author's
knowledge, it was the first conference of its kind
to specifically target women scientists at HBCUs.

2005   Condoleezza Rice becomes the first black woman
to serve as secretary of state for the United
States.

A resolution is proposed for the contributions of
African American women in science by congress-
woman Eddie Bernice Johnson on April 26.

Table 1. Total number of doctorates awarded to U.S. citizens in science and engineering disciplines by race and gender in 2000

	Science (All Disciplines)	Engineering (All Disciplines)
*Total number of doctorates	14,508	2,556
Black Women	379	25
Black Men	266	58
White Women	5,177	329
White Men	6,360	1,549

*Includes all races and gender of U.S. citizens and permanent residents.

Data were compiled from summary reports from the National Science Foundation.

Table 2. Number of bachelor's degrees earned by black women which are conferred by historically black colleges and universities)

	1994	1995	1996	1997	1998
Engineering	265	298	350	371	350
Mathematics	268	262	237	255	229
Computer Sciences	523	556	537	456	466
Biological Sciences	863	972	1,146	1,298	1,345
Agricultural Sciences	65	65	85	104	121

Data were compiled from a summary report provided by the National Science Foundation.

Table 3. Number of bachelor's degree earned by black men which were conferred by historically black colleges and universities.

	1994	1995	1996	1997	1998
Engineering	425	500	544	534	493
Mathematics	181	204	206	223	213
Computer Sciences	381	411	442	378	438
Biological Sciences	334	385	482	543	573
Agricultural Sciences	69	58	112	112	99

Data were compiled from summary reports from the National Science Foundation.

Table 4. Total number of engineering doctorates awarded to women who are U.S. citizens, 1996–2000

	1996	1997	1998	1999	2000
Black Women	19	22	21	23	25
White Women	324	312	278	332	314

Data were compiled from the summary report 2000 provided by the National Science Foundation.

HATTIE CARWELL
Civil Rights for All

MOST PEOPLE RECOGNIZE HATTIE CARWELL for her book *Blacks in Science: From Astrophysicist to Zoologist* (1977). However, Carwell has worked tirelessly to promote and encourage people of color and women in science and technology throughout the world.

Carwell grew up in Ashland, Virginia, during the 1950s and 1960s. Raised by her aunt and uncle, she grew up in an all-black community, where she was nurtured and encouraged to do well in school. Hattie recalls becoming interested in science in junior high school, where she took accelerated science and math tests. Because the teachers were a part of and lived within the black community, they knew the students inside and outside of the classroom. This nurturing community spirit proved invaluable in boosting Hattie's confidence and fostering her decision to pursue science and technology.

Hattie Carwell, one of the first black female health physicists, also participates in community activism.

After graduating as valedictorian of her class, Hattie went to Bennett College in Greensboro, North Carolina. Bennett College, one of the only two still-existing historically black women's colleges in the U.S., was founded in 1873 as a coeducational school in the basement of Saint Matthew's Methodist Episcopal Church. In 1926, Bennett became a college for women, primarily as a result of the Women's Home Missionary Society's desire to enlarge its educational programs to include African American women. Hattie chose Bennett partly because of its great history, and because she received a scholarship and wanted to leave Virginia. She further recounts: "I wanted to go to a place that offered not only an opportunity to be trained in a particular field but also an environment that offered me other cultural opportunities." Bennett was the place that offered both.

Hattie had grown up in an environment where she had been motivated to study science, and her college experiences enhanced her love for science. After receiving her B.S., she journeyed to Rutgers University, where she received her M.S. in health physics. Rutgers was an ideal place to study because it allowed her to be close to her family.

Her first position was as a research assistant to a radiation biologist at Thomas Jefferson University hospital in Philadelphia in 1971. She stayed there only for a year, then she began working at the Atomic Energy Commission in the Brookhaven area office in New York. She worked at the Energy Research Administration, which later became known as the Department of Energy (DOE). She got an unusual opportunity to work part-time in New York and in the San Francisco office. This new arrangement was appealing and worked well with Hattie's desire to live on the West Coast.

Carwell has held a number of positions at DOE, from being a senior health physicist's senior facility operations engineer to becoming operations lead officer at the Berkeley site office. Hattie faced challenges in moving beyond the glass ceiling but attained several senior management positions, primarily in the safety management and in the proper maintenance and management of waste facilities. She has continued her studies in biophysics at the University of California at Berkeley and received an honorary doctorate from Bennett College in 1993. She is a strong advocate of being involved with the black community both locally and globally.

When did you first become interested in science?

Carwell: I think it was probably around the seventh or eighth grade. I participated in a science competition every year.

You mean like a science fair?

Carwell: No, it wasn't a science fair. The teachers worked with students at an accelerated pace on science and math. Every spring the students would sit and take a test on their math and science skills. That happened every year until the twelfth grade. It was quite exciting and fun for me.

Was there any particular person that encouraged you to select science as a career?

Carwell: No, it was just the process that I described earlier. Each year it would be a different teacher who would do the testing. Although I did have a Sunday school teacher who was also a science teacher. She had a very positive response to my interest in science and nontraditional areas. I automatically took the college preparatory courses and I graduated valedictorian from my high school. I don't think there was any high school counselor that didn't encourage me to take preparatory courses. I loved taking all the science and college preparatory classes.

What role did your family play in your selecting science as a possible career?

Carwell: I would say that my family played a minimal role, other than encouraging me to do well in school. I was free to decide what to do. I was raised by my aunt and uncle. They were not professionals and they did not voice any particular opinion about what I studied. They just wanted me to do well.

Where did you grow up, and did geographical location have anything to do with your selection of science as a career?

Carwell: I grew up in Virginia, and I don't think geographical location had anything to do with it. I think, maybe, it had to do with the community spirit. I grew up in a small town called Ashland. It was an all-black town. The teachers were very supportive and interested in their students.

The teachers were in the community and they interacted with you in school and church. They knew your family, so this fostered a strong relationship.

Do you feel that race and gender have played a particular role in how you are viewed as a scientist?

Carwell: Most definitely. Unfortunately, over the course of my career, race and gender issues have not changed dramatically.

Can you cite some specific examples of how race may have been a factor?

Carwell: I will use an example from my workplace to answer that question. I have followed the careers of people who worked in a similar situation over the years. In most work situations, opportunity is based on assignments, and if you are not given appropriate assignments then you cannot advance. In other words, you are tracked or groomed for some of the choice positions. If you are not getting the right experience, then you are automatically going to be eliminated from being competitive for those positions. I see people just assume that because the worker or employee is a white Anglo-Saxon male that he should get certain experiences, whether he has taken the initiative or not. However, minorities have to be assertive and creative to seek out opportunities to gain the experience that they need to move ahead. It is not an automatic assumption for minorities. I think those assumptions that are made are racially based and can play a role in career development.

Are there specific examples of how gender has played a role?

Carwell: It is difficult to separate out the race and gender issues. I think when you first start, it's more race than gender. As you become more senior and begin competing for the higher-level jobs, then gender may become more of a factor. Males want those positions as much as you do. There are fewer senior-level positions.

You spent five years of your career in Europe. You worked on loan at the International Atomic Energy Agency in Vienna, Austria. How was that experience, and did your race and gender play a role during that time or upon your return to the U.S.?

Carwell: I would say working in Europe was one of the best times in my life. I actually sought out that experience. It was one of the best work experiences that I ever had, especially reflecting back on my career at this time. From a career standpoint, it was a good opportunity to really develop professionally, but I consciously decided to come back to the States.

In terms of being involved in the black community, the experience was lacking. I remained involved to the extent that I could. For example, I was still involved in the technical organizations that were geared towards African Americans like the National Technical Association and the Northern California Council of Black Professional Engineers. I even started to write another book when I was there, but I missed being deeply involved in and connected to my community. I also felt that if I was going to have a career back in the U.S. it would probably mean that I shouldn't stay abroad too long. In some ways, my being abroad impacted my career somewhat negatively.

Actually, it gets back to your question. I felt there were a lot of petty jealousies from some of my white male colleagues. I worked with the Department of Energy, which is a government agency. As a government agency, the office was supposed to keep employees in the U.S. and abroad abreast of career opportunities within the agency and one's division. When I returned from Europe, there were new people who had been brought into my division. Basically, they were unqualified and very racist. They did not inform me of the promotion and career opportunities. Unfortunately, one particular person from this group became my boss, and he had less formal education in my particular area. He felt very threatened and there were several negative things that were done that impacted my career. He was very overt and clearly showed a preference for selecting other white males, regardless of qualifications. Fortunately, an internal investigation was done and I became aware of some of the racist things that were being done. That position was not filled.

Subsequently, I did apply for a promotion outside of that division and I received that one. On the surface, it may appear that what was happening to me was what might happen to anyone else or any woman, but when I dug deeper it was generally agreed among other supervisors that what I perceived to be happening was true.

> *That is, because of your race and gender you were being denied an opportunity to advance in your career.*

Carwell: Yes, I was more qualified than those selected for the promotion. Like I said, as you move up, it is sometimes harder to distinguish between race and gender, but my former boss was clearly being a racist.

> *What about your experiences in Europe specifically?*

Carwell: I have had positive experiences in my career, especially before I left for Europe. In Europe, I had a continuation of positive work experiences. I worked in job situations there that weren't necessarily based on me as an individual. My experience had more to do with what country I was from. Each country had a certain number of positions that were more related to politics than anything. I also think you find less of a problem related to race and gender in the early stages of your career. Of course, each person has his or her own experiences. When I came back to the U.S., I began to experience problems related to race and gender. I think it partly had to do with competing for senior-level positions and just the pettiness and insecurity of other people who do not understand or relate to your culture or you as an individual.

Are there other factors other than race or gender that affect how you are viewed as a scientist?

Carwell: Well, I would label it as culture or style. Women tend to have a little different style than men. Also, your cultural background may influence your style. There's a tendency for black women to have a straightforward style. Many times they have been reared in an environment where honesty has been encouraged. In many corporate environments, that honesty is not necessarily a valued trait. That is, knowing how to play the game in the workplace is considered far more valuable than giving your honest opinion about something. People tend not to advance as fast with that kind of an approach. Even if you have the appropriate experience, how people perceive you as a team player becomes far more valuable than your ability or experience.

Do you feel that the civil rights movement played an important role in your success as a scientist or your career?

Carwell: I don't think it did in my selection of science. But once I was in the career, I think that the civil rights movement enhanced my opportunities.

Do you feel the women's movement played a role in your selection of science or your success in science?

Carwell: I became aware of the women's movement after I had chosen a science career. I don't think it played a role because I grew up in a small town in Virginia in an environment where people did not tell me what I could not do.

I was captivated by science. I was surprised that more of my classmates did not go into science. I did have women as math, chemistry, and physics teachers. I was not aware that there was a gender problem when I was growing up. Everyone was encouraged to do his or her best at the segregated high school that I attended.

Are you saying that your early experiences tracked you into science?

Carwell: Yes. I did all the science courses. My family wanted me to go to college. I did not necessarily know how I was going to pay for college, but I was surrounded by people that wanted me to do well. There was only one person who got upset when I chose to do science. My uncle thought that I should be a secretary or a teacher, and I think that his negativity simply reflected his lack of understanding of career options for women.

How do you feel other scientists can be supportive of black women scientists, and what responsibility does a black woman scientist have in her own success?

Carwell: Ultimately, every individual is responsible for him or herself. You just have to work at knowing what you do well and what you do best. You have to set goals and research career pathways that may be best for you. No one is taught a career path, but one has to invest time in learning and understanding how to reach certain goals in a particular career. One should be willing to take risks. That is, you may not take a direct pathway to the career that you end up in but you have to be open. You need to share your ideas with someone that you trust and can give you an honest assessment of where you should be in a science career. Others can be very helpful and supportive, but you bear the ultimate responsibility. We black women scientists should be willing to share our experiences with younger women. There is a lot to be gained from our support of each other that's very valuable.

> *How have marriage and family played a role in your life as a scientist?*

Carwell: I am not married. So marriage and family have not played a role in my life as a scientist.

> *How has being single played a role in your life?*

Carwell: Well, I guess I've had a certain amount of freedom in my decision-making process (based on my observation of married people). I have been able to choose a variety of experiences and activities to become involved in that my married counterparts have had to make a different decision about. I spend a lot of time with science-related activities and youth in the community. As you know, I have also documented the achievements of blacks in science. I have written one book on the subject and several articles.

> *What led you to write about other scientists?*

Carwell: I am a person who has always had varied interests. I heard someone give a talk on black scientists and their achievements. It was very stimulating, and I tried to find more information on black scientists, but it was very difficult. There were a number of books that just repeated the same information. Then I started to go through periodicals to find information. I found that it was very satisfying to find more information through that route, and so I thought I would share this information in a book form.

> *I can relate to that lack of information. That's why I am doing this book on black women scientists. I noticed that your book did not have a lot on black women scientists, but I am aware of an article that you wrote on African American women in science. Was there a particular*

reason as to why you focus more on the achievements of black males? I am guessing that you lacked information on women at the time that you wrote your book.

Carwell: I think my book is reflective of the limited written material available on women. Also, the format of the book restricted including some women, because it covered all fields, from astrophysicist to zoologist. Many of the early black women in science were biologists or mathematicians. In recent years, we find more women in engineering, physics, computer science, etc.

Because of the long, rigorous hours in science, some young women feel that they have to choose between a career and marriage or family life. What advice do you offer a young woman with these concerns?

Carwell: Well, that may be the case, but I think there are so many science-related jobs that one can choose that don't require overtime. If you are in research, then you will probably have to work overtime. I think it depends on how involved you want to be in the job.

I also volunteer a lot of my time. I think there are a lot of ways to contribute to science because there's more flexibility in the workplace. I think that young women have to understand that if you choose a certain path like academic research, it might require long hours. When I worked in Europe, I had to travel 60 percent of the time, and there were men that were married and doing the same thing. You have to have a supportive spouse or tailor your career path accordingly.

What do you think black women can do to become more visible?

Carwell: I don't think black women are invisible, unless you think black people (in general) are invisible. Our numbers are significant.

No, that is not what I mean. I think you will rarely see the contribution of black women scientists shown or talked about in a real, significant way. Yes, many people will know or be familiar with George Washington Carver or maybe Ernest Just. Now, a few people might know Mae Jemison.

Carwell: Overall, yes, that is true. The contributions that we have made are invisible. You have to do more than just your job. You have to extend yourself beyond just what it takes to get a paycheck. Science and technology affect the way we live. It is important to the development of African Americans. I extend my science work to the black Diaspora. I extend it to the development of people of African descent throughout the world. There is so much that we can do to help the nonscientist to understand the role of science and technology in improving the quality of life. It's important to take the time to share the information in your community.

> *Do you think professional organizations have a particular role in promoting or serving women of color?*

Carwell: My experience is that when the organizations speak about women, they are talking about white women. I know that they are pushing for more women and trying to improve their status.

> *What do you mean? Do you mean serving on women's committees in various professional organizations or women's organizations?*

Carwell: Both. Black women are not focused upon in either situation. Just take a look at the women's committees and the organizations for women. They are mostly white women. For example, the DOE has an annual review of the status of women at national laboratories. I am the contact at this office for our group. I know there are a lot of programs for women and I look at that as one resource for getting the message out. The perceptions are that these programs are for white women, but they are hard-pressed to turn away black women who get involved.

In the African American community, you just work as a unit. For example, I am involved in a national scholarship program and I try to make sure that there is a balance of both male and female applications. For conferences that I am coordinating, I always make sure that we have at least one woman as a major speaker. I am willing to make the extra effort to find or identify the woman speaker, if necessary.

> *So are you saying that white women only seek white women, which adversely affects black women's participation in science organizations? You seem to imply that that is the case, even though you would try to select personally a broad perspective.*

Carwell: I mean that the women's movement tends to address issues from a white woman's perspective. Groups promote the interests of the membership. For example, the Society of Women Engineers is composed primarily of white women. They recruit the women engineers that they socialize with. Most of their social friends are not black women. In fairness to them, there is no effort to exclude black women, but there is no effort to recruit them. Black women are not knocking on the door to be included.

Black women engineers and scientists tend to support black efforts to increase numbers of both groups. In other words, organizations that encourage the development of both women and minorities will have a stronger attraction and affinity to black women. We support a more inclusive approach to organizations.

How do you envision yourself in ten years?

Carwell: In ten years, I will probably be retired. I would really like to build a foundation and do some consulting with developing countries. I met the president of Uganda in 1994. I've worked with the African Regional Technology Center in Dakar. I have a project in Ghana. I am currently developing an Internet project called Global Village Concepts. I think I will always have an interest in writing about African American contributions to science. I am currently working on a booklet that I am hoping to expand into a book about Dr. Warren Henry [now deceased]. He is probably the most eminent black scientist that is still living. He worked on radar in World War II at the MIT radiation lab. He invented the video amplifiers to prove signaling capability of the radar scope. In fact, he had eight inventions. After the war, he worked at the Naval Research Laboratory and became involved in research on paramagnetism. His research is so fundamental that it is used in our basic physics textbook. This textbook is used in England and Russia. There are very few people who know about him.

I think that is wonderful. What do you think your greatest contribution to science will be?

Carwell: I used to say my book and my writings about blacks in science. Now I think the scholarship fund that I chair has been very significant. The scholarship fund came into existence in 1983 during the Reagan administration, when they were making it difficult for African Americans to afford an education. A group of professionals decided to get together to develop a fund for African American students. Each professional agreed to contribute $1,000 a year to the fund. In 15 years we have given over $150,000 worth of scholarships, even though we have a small number of members (less than 25). Our major efforts are in selecting the students and supporting them until they graduate. Our goal was to demonstrate that we could and would give back to our own black community.

Is there anything that you would do differently in your career?

Carwell: Well, I don't know whether I would or should have. I probably should have stayed in Europe to advance my career. I probably should not have gone back to the same office once I returned from Europe. Personally, I wanted to contribute to the advancement of black people. I was very comfortable in Europe but I was too far removed. By returning to America, I could work in science and deal with the issues of African Americans that were dear to my heart. Now that I think about it, I probably would not change a thing.

What advice would you offer to young black women?

Carwell: If they enjoy science, then they should pursue it. Science is not for everyone, even when you have the skill and ability to do it well. The most important thing is to position yourself to get what you really want in life and really prepare yourself. It is also important not to be afraid to go in one direction and then change if it's not what you really want in life. You are still black and female (in this country), and no matter what decision you make, it is not easy. However, science is no harder than any other discipline. But you have to make up your mind and put your mind and best efforts to the subject. Preparation is a major key to success.

Finally, don't isolate yourself. Share your dreams and aspirations with others. Share your negative and positive experiences with others. By doing so, you will know that you are not alone and not the only woman who has had some negative experiences. There's a way to turn the negative into the positive. I encourage young women to pursue their dreams and work hard.

Interview Date: August 1998.

Selected Publications and Research Activities

Blacks in Science: From Astrophysicist to Zoologist. New York: Exposition Press, 1997. Currently in 3rd printing.

Articles published: More than fifty in journals, magazines, and newsletters (technical and on social issues related to technology).

Member of the Editorial Committee for the National Technical Association Journal, 1980–1994.

Editor, The Real McCoy Newsletter, 1976–1980; Staff member, 1985–1990.

Operations Lead at Berkeley Site Office, January, 1994– .

Program Manager for High Energy and Nuclear Programs with the U.S. Department of Energy, San Francisco Operations Office, Oakland, CA, 1990–1992.

Assistant Environmental Survey Team Leader, DOE headquarters, 1987.

Senior Health Physicist for the Atomic Energy Commission, Brookhaven Area Office, Energy Research Administration and U.S. Department of Energy, San Francisco Operations Office, 1972–1989.

Yvonne Young Clark

Still Going Strong

AT 74, Y. Y. CLARK (AS SHE OFTEN REFERS TO HERSELF) is still going strong in the classroom. She's still teaching one of the largest freshman engineering classes at Tennessee State University in Nashville, where she began her academic career fifty years ago. Like many women in science, her interests began early and were nurtured by loving and supportive parents. Yvonne recalls, "When I was a child you could put something in front of me and say, 'Put this together,' and I could do it because I loved doing things like that."

Clark was born in Houston, Texas, in 1929 and spent her childhood in Louisville, Kentucky. Her parents were educated at the premier black colleges in Nashville. Her father was a physician and surgeon, and her mother was a librarian. Both were graduates of Fisk University, a liberal arts college that was established in 1866 to educate the newly freed slaves. Her father received his medical

Yvonne Young Clark was the first black woman in the United States to receive a B.S. degree in mechanical engineering, at Howard University in 1952.

training across the street at Meharry Medical College, one of the few medical schools established to educate black doctors.

Although Yvonne came from an educated household, she did not necessarily see her parents as role models for her career choices. Clearly, she was exposed to all the opportunities that a middle-class black family could provide in the 1930s and 1940s. In school, she had a well-rounded background. She took Latin, Spanish, and three years of science courses. She even took an aeronautics class where she built airplanes and crash-landed them from the fire escape at her high school. By any standards, she seemed well-prepared for college matriculation and a major in science or mathematics.

In 1947 Yvonne broke family tradition by deciding to enroll in Howard University instead of Fisk University. At Howard University, she found a nurturing and supportive environment. She recounts, "At Howard they told me that I might have a hard time, but if I was prepared nobody could stop me." And nobody did stop her. She did well in her studies and was happy about her decision. She graduated in 1952—the first black woman to receive a B.S. degree in mechanical engineering.

Yvonne, like so many college seniors, interviewed with companies that came to campus during career day. She encountered one recruiter who tried in every way to discourage her. The recruiter told her "that his company brought their engineers in at the bottom and that they would have to work their way up." Yvonne said that she would not have a problem with this arrangement, but she quickly recognized that the recruiter did not want to hire her because of her gender. Clearly, the company recruited students from Howard, but she was a woman and he did not want to hire her. This blatant discrimination did not deter Yvonne. She finally landed her first job, at Frankford Arsenal-Gage Laboratories in Philadelphia, designing gages and making final drawings for their products. Three years later, she accepted her second position, at RCA's Tube Division in New Jersey.

Love played a role in Yvonne's next move. On summer visits to her cousins in Nashville, she met her future husband, Bill Clark, Jr. He was an instructor of biochemistry at Meharry Medical College. After some courting time, she agreed to leave New Jersey and come to Nashville to be with Bill.

That was the beginning of her long history in Tennessee. Yvonne continued as a pioneer in so many areas of engineering. She was the first black woman to become a professional engineer in Nashville and the first female faculty member in the College of Engineering and Technology at Tennessee State University. She was also the first female to head the engineering department, where she began as instructor and worked her way up to associate professor. In the summers, she continued to work as a professional engineer at various places in the South, including NASA in Huntsville and Houston and Westinghouse's Defense and Space Center in Baltimore.

In 1970 Clark was in the vanguard again when she became the first black woman to receive a master's degree in industrial engineering, at Vanderbilt University in Nashville. Although Yvonne got involved in an academic career because of the lack of industrial or corporate opportunities when she returned to the South in 1956, she has never regretted the hand that life dealt her. In fact, she has gone on to become one of the trailblazers in engineering education for women and minorities. She was one of the early members of the Society of Women Engineers. She is also a member of the National Education Association, the Tennessee Education Association, the American Society of Engineering Education, and the American Society of Mechanical Engineers.

In the larger Nashville community, Clark has been involved in several service organizations, including the Boy Scouts and the Delta Sigma Theta Sorority, Inc.

Because of her example, some of the boys that she led in the Boy Scouts chose to become engineers. These experiences make it easy to see why Y. Y. Clark is a builder of minds and lives. She has built and paved the road for more than four generations and is still going strong.

How did you first become interested in science?

Clark: I always liked to build things as a child. I had an Erector set and I fixed things around the house. I loved to do it. So I guess I had the engineering bug in me early on.

Did your parents influence you early on to pursue science or engineering?

Clark: Well, they were supportive. My father was a physician and my mother was a teacher and librarian. I grew up in Louisville, Kentucky.

Was there any particular person who influenced you or encouraged you in science?

Clark: No, there was no particular person who influenced me in selecting science as a career. I really didn't have role models in science when I was growing up. My parents were professionals in the black community, but I wasn't really interested in the fields that they had chosen.

What role did your teachers or community play in your selection of a career? Did geographical location have anything to do with your selection or interest in science or engineering?

Clark: Neither the teachers nor the community supported me in my career choice. At that time, women were limited by the narrowly defined gender roles and sexism.

Geographical location also didn't have anything to do with my selection of engineering, but it did affect where I could go to college. Most of my family went to Fisk University, but Fisk didn't offer any courses in engineering; so I broke with family tradition. I could not go to the University of Louisville because I was south of the Mason-Dixon line, so in that sense, location played a part in where I pursued my science and engineering major. But it had nothing to do with my interest in the field.

Do you feel race or gender has played a particular role in how you are viewed as a scientist?

Clark: I think both have probably played a role, but I have never let it stop me from succeeding. When I was in high school, I wasn't allowed to take mechanical drawing. When I came back to visit my high school after my freshman year at Howard, I went to see my teacher, Mr. Adams. I told him, "If a girl wants to take this course, you'd better let her have it." I had entered Howard unprepared because of him. I would have done better if I had had that class in high school.

Did your actions change the school's or his policy?

Clark: Yes, girls now take the class. That's pretty much how it was at that time. When I was at Howard, I was the only woman in mechanical engineering and the first woman to get a degree in it. There were females in electrical and chemical engineering because the electrical and chemical areas/departments didn't have the same stigma of "getting dirty" as mechanical engineering had. Therefore, I was the only female in the School of Engineering during my matriculation at Howard University.

At Howard, the mechanical engineering faculty encouraged me by not discouraging me. So it was a matter of my own self-determination and motivation to do what I wanted to do.

What about race and gender issues in your professional life?

Clark: Well, I really don't think those issues affected my success, but I can relate a couple of examples in reference to your question. I interviewed with several companies that came to the campus. One company told me that their engineers had to start at the bottom level so that they would know how the machinery worked and then work their way up. I said that would be okay by me. Then he said I didn't have the muscles to break down the machinery. I said, "That's not a problem, but don't use my lack of strength as an excuse not to hire me. Just tell the truth and say that it is because I am a female." Of course, others, like the Navy at that time, did have the nerve to say that to me.

What about issues related to race? I would think that might have been an issue during that time, in particular.

Clark: In 1956, when I came back south to Nashville, I had been told by the Ford Glass plant that they were not interested in hiring me. I think that was related to my race, but I got a job teaching. In 1970, when I was working on my master's degree in engineering management at Vanderbilt University, I needed an industry job to complete my data for my master's thesis. The supervisor of engineering at the plant also served on the advisory committee at Vanderbilt University. He decided to check with the company and he was able to find me a position there. I became the first female engineer at the plant and I broke the ice.

While I was at Ford, I was invited to a luncheon. I was the only black person at this event. One of the white men got up and left the table; everyone was upset because they knew why he left. Personally, I was not troubled. I said, "Oh, he's going to miss a good meal because the steak looked good." I thought, Why should I be upset because he is a racist or has a problem for whatever reason? It was his problem and not mine. So I have experienced race and gender incidents but I try to remember that that's the other person's problem and not mine. This is where I think I have been really successful.

Are there any other incidents related to gender that you can recall?

Clark: Between 1970 and 1972, having a Ph.D. became the requirement for the professorship at TSU, or maybe at all Tennessee Board of Regent Schools, I do not know. Before 1970, I placed an EEO complaint against Tennessee A & I State University because, in my opinion, I was more qualified than some of the men who were given the rank of full professor in the School of Engineering and Technology, and I could not progress past the associate professor rank. I also believed that since I wore a skirt I would never be a full professor. I have been in the rank of associate professor since 1962. I never wanted the Ph.D., but I thought if I pursued my professional license, it would mean something within the engineering profession, although not in academics. I think I was right. This was my thought process, and, for me, it worked.

How long have you been a professional engineer?

Clark: I went before the Architecture and Engineering Board in 1960 and qualified for my professional license. I was then able to practice in Tennessee. This allowed me, if I wished to, to augment my salary. I received my certificate in the mail and later found out that all new registered engineers were invited to a monthly luncheon meeting and introduced to the Tennessee Society of Professional Engineers members. I was invited to join the state society in 1964. Since that time, I have been both a financial and an active member. In February 2001, I received the Most Distinguished Service Award at the All-Engineers Banquet, and in 2002, I introduced the recipient of this award.

Are there factors other than race or gender (like class, age, etc.) that influence how you are viewed as a scientist or engineer?

Clark: Race and gender strongly influence the ways in which the black female scientist/engineer is viewed. In fact, anything else is secondary or even minor.

Do you feel that the civil rights movement played a role in your success in your career?

Clark: No, I was already there. The push for equal opportunities in the workplace for black people and women had already been accomplished for me, the individual.

Would you say that the women's movement did not play a role in your career?

Clark: The women's movement did not help me either. I was already there.

How do you think other scientists can be more supportive of black women scientists? What responsibility does the black woman scientist or engineer have in her own success?

Clark: Black women scientists need to participate in professional organizations and to network and form mentoring relationships. It is incumbent upon us to make sure we are active in our national and local organizations. I have been a member of the Society of Women Engineers since 1952. I am also in the National Education Association, the Tennessee Education Association, the National Society of Professional Engineers, the American Society of Engineering Education, the Tennessee Society of Professional Engineers, the American Society of Heating, and the Refrigerating and Air-Conditioning Engineers, to name a few. Once we are active in our profession then we can get and give some of the support that we need.

How has marriage and family played a role in your life as a scientist?

Clark: I am a widow now, but I had no problem in combining my career with my family responsibilities. My son was born in 1956 and my daughter was born in 1968. I was lucky enough to have leave for the fall quarter when my son was born. I had about six weeks to eight weeks before the fall quarter began when my daughter was born. My husband and a housekeeper helped take care of the children. I was really lucky. These days I get to enjoy my grandkids.

Are there any particular strategies that you would recommend for someone who is balancing a science career and family?

Clark: The balancing of a science career and family is dependent on the choice of a mate. His maturity, understanding, and willingness to plan together are key to successful careers. The basis for a lasting relationship is one in which the mates have the same goal, a desire to aid the other in being all she/he can be. Obviously, this may mean times when one has to relocate for the other, or one makes more money, etc., but it is agreed upon.

What can we do as black women to change our invisibility and image in science? How do we become more visible?

Clark: I think we have to be active participants in our professions at all levels. We have to be proactive: network and participate in organizations which are outside our science profession but which allow us to impact others. We can't sit and wait or allow another person's perception of us to deter us from our goals. As I mentioned earlier, I never allowed anyone else's problems to be mine.

Do you think professional organizations have a particular role in promoting the visibility of black women in science? If so, what do you think that role should be and why?

Clark: Yes, it allows us to help develop sensitivity/awareness to racial and gender issues at decision-making levels. Additionally, it allows us to model our expertise and leadership skills and act in supportive ways.

How do you envision your future as a scientist? Where do you see yourself in ten years?

Clark: As long as I am a practicing engineer and teacher, I envision myself encouraging, educating, and networking for young engineers. Thus I see myself as increasing the engineering workforce.

What do you think your greatest contribution will be to science?

Clark: I really enjoy teaching. I love to see the progress of my students and their accomplishments. There is no greater joy than when they return and they tell me that they have gotten their master's or Ph.D. or they have passed the professional engineer's exam. This is what my career has been all about. This is meaningful in my life.

To me, the fact that I have opened doors and continue to be instrumental in opening doors which lead to the advancement of black engineers, especially black women engineers, is my greatest contribution.

What advice would you offer to young women, especially young black women?

Clark: I would say that young women, especially young black women, should not allow gender limitations to deter their career choices, to be certain that they develop their skills and knowledge base, and that they begin early participating in professional organizations and networking.

Interview Date: November 1997.

Selected Publications and Research Activities

Thesis: Designing Procedures for Materials Flow Management in Major Rebuild Projects in the Glass Industry (Vanderbilt University, 1972).

Women in Engineering Education. In IEEE-Education. Co-authored with Lilia Ann Abron.

Development of the Moments and Moments of Momentum for the Six Degrees of Freedom of a Rigid Body with Respect to the Body Axes and Earth Fixed Axis. (U) Army Msl Cmd Rs-Tr-62–4, March 5, 1963, 22.

Cooperative Development of a Bibliographic Data Base for Ocean Thermal Energy Conversion System Technology. Funded by the Department of Energy (1985).

Energy Usage Monitoring of Residential and Commercial Structures. Funded by the Department of Energy (1984–1987).

Experimental Evaluation of the Performance of Alternative Refrigerants in Heat Pump Cycles, Department of Energy. (1987–1997).

ANNA J. COBLE

To the Beat of Her Own Drum

ANNA J. COBLE'S INTEREST IN MATH AND SCIENCE was cultivated early by her father, who worked at the local college, across the street from the family home. Born and raised in Raleigh, North Carolina, Anna's early interest in math and science was further encouraged by her high-school teachers. She recalls having "very good, strong women teachers who were using their minds in the sciences." After high school, Anna attended Howard University and received her B.S. degree in mathematics in 1958 and her master's degree in physics in 1961. She then returned to North Carolina and taught in the physics department at North Carolina Agricultural and Technological University.

After about four years in teaching, Anna felt it was time for a change in her life. She decided to return to graduate school to work on her doctorate. In-

Anna J. Coble, the first black woman to achieve a doctorate in biophysics, is devoted to women's and minorities' advancement in science.

terestingly, Anna's choice for graduate school, the University of Illinois, can be traced back to her godfather's influence. He had graduated from that university in the 1940s, and she had heard positive things about the program at Illinois. By that time, she had become more interested in applied physics than theoretical physics, and the biophysics program at Illinois was in the physiology department and was a good fit for Anna's future goals. Because Anna was not a typical graduate student, but an older and more seasoned person, she integrated very easily into the program and curriculum at Illinois. She became involved in the university's recruitment of, and efforts to retain, African American students. Like many majority institutions that attempt to recruit minority students, the University of Illinois strategies were fairly typical and unsuccessful. However, having someone like Anna Coble as a student and staff member led to improvements in recruitment opportunities. The university was tapping such cities as East Saint Louis for the black undergraduate population. Anna recounts,

> They were taking these kids from East Saint Louis who were not ready for the kind of grind at a major university. Their approach was not working. They [the university] brought in about 200–600 black students with no preparation for campus life and the results were disastrous.

This was the beginning of Anna's advocacy for minority students and women at the University of Illinois and she has continued for almost four decades on their behalf.

Before she finished writing her dissertation, Anna accepted a position at Howard University in 1971. It was like coming home. It took her about two years more to finish all her requirements for the Ph.D., which was awarded in 1973. In addition to her previous teaching experience, Anna had gained her share of research experience in the Midwest, working with the well-known environmentalist Barry Commoner at Washington University in St. Louis for two years. With her teaching and research experience, Howard seemed like a good opportunity.

Anna had also made history by being the first black female in her department at the University of Illinois to complete a Ph.D. in that program and by being the first black female to be hired in physics at Howard University. Indeed, according to records searched, she was the first black woman to receive a doctorate in biophysics.

Anna has been a faculty member at Howard University for over thirty years. She has worked tirelessly on and off campus to promote diversity in science. She has served as the local and national chairperson for the Network of Minority Women in Science. She had taught courses for science teachers as well as the AP physics for senior high school teachers. Her professional activities are numerous and include being a Minority Access to Research Careers faculty fellow at the National Institutes of Health from 1979 to 1980 and organizing the Science Discovery Day for the Washington area. This program has become one of the most successful and popular for attracting junior high school students interested in science. She has served on several National Science Foundation panels as well as being a panelist for the MacArthur grants administered in New Jersey.

Her passion for helping others has extended into her community, where she is a member of the board of directors for the Ionia Whipper Home, a shelter for abused, abandoned, and neglected teenage girls. With all of these activities, one might think Anna doesn't have time for any other social or family activities. Not so. In addition to her great talent as a physicist, she has a beautiful singing voice and has been a featured soloist with the University of Maryland Choir, the Howard University Choir, and the Evelyn White Choral Ensemble. Unlike many women of her generation, Anna married later in life, at the "ripe young age" of 49. She teasingly says, "I wasn't looking for a husband, but he came by." She encourages young women to pursue their own dreams and not to get caught up in someone else's definition of them. Anna is truly a role model for young scientists, especially for women and minorities. She doesn't get caught up in titles or accolades. When asked why she has not attained the rank of full professor, she replies, "I haven't bothered and it is not important. I am an advocate for our students and women." This is a woman who simply marches to the beat of her own drum.

Was there a particular teacher or family member who influenced your choice of selecting science?

Coble: My father taught math and physical science at St. Augustine College, so I always had a family member interested in math and science. One high school teacher, Susie Perry, had a master's degree in chemistry, and Virginia Newton, a math teacher, had a master's degree in mathematics rather than in education. These were very good, strong women teachers who were using their minds in the sciences. By the time I got out of high school, science looked like it would be a continuing challenge, one that I eagerly took on.

Where did you grow up?

Coble: I was born and raised in Raleigh, North Carolina, and I lived across the street from St. Augustine College. My bachelor's was earned in mathematics in 1958, and my master's was earned in physics from Howard University in 1961. I went to teach right after I earned my master's degree, back at North Carolina A & T University in Greensboro.

What were your student days like at Howard?

Coble: The physics department and most of the science departments at Howard in the 1950s were very encouraging of women. Mordecai Johnson, president at that time, really liked to see women in the sciences—particularly if they could present themselves well and talk about things that he just had no chance of understanding! He was impressed with that sort of thing. That was generally the type of environment we had at that time. Now, there were those who would make comments. I remember my department chairman openly made some comment during a class discussion about not telling the girls about a code for a particular device. And I was the only female in the class. I said at that time that it was not worth dealing with. I continued my studies, but under certain circumstances there were those who would not encourage women. My experiences were mostly very encouraging as a student.

What happened after you received your doctorate?

Coble: When I was finishing my dissertation, I had reams and reams of data on strips of tape that needed to be analyzed. I went to teach at Howard University in 1971 before finishing the data analysis. I have a Ph.D. in biophysics, but I prefer teaching in a physics department.

Tell me about your experiences working on the doctorate.

Coble: I went to the University of Illinois and received my Ph.D. in biophysics in 1973. My experiences were different from a lot of people's. I was an older student returning to graduate school. Remember, I had already worked four years in North Carolina. The whole department at Illinois was supportive. I couldn't say something to one person about somebody else without it going through everyone in the department. That made me a little careful, but it also told me about their concern for me and my experience. That was an important reflection on the department, and they were willing to change and hear my point of view. For myself, I did not feel restrained, as other students did. My advisor invited me to all the department parties and I had dinner at different faculty homes. When I was there during the Thanksgiving holiday,

I had a special invitation to his home. There was this kind of cordiality. This kind of treatment really made a tremendous difference in my experience at that institution.

Were you the only black in your department?

Coble: Yes, I was the only black graduate student in the biophysics program at that time. This was in the mid-1960s. There was another black graduate student in the department. He got his Ph.D. from Illinois. From there, he went to Iowa and earned an M.D. and a J.D. His whole interest was in increasing the number of black doctorates in the medical sciences and in improving the programs in which they were educated. I don't know how much of that he's doing in his work now.

He sounds like an interesting person. Tell me about the black undergraduate experience. I understand that you had a big part in establishing programs for them.

Coble: Yes, our experience as graduate students at Illinois was great, because we would get together and plan activities. At that time, there were about sixty-five of us. The university was trying to increase the number of black undergraduates, but they did it by taking kids from East Saint Louis and things like that. These kids were not ready for the kind of grind at a major university. When they brought the black kids to campus before the rest of the freshmen, they put them in the newest dorms, but when the fall classes started, they moved them to these old dormitories where all freshmen usually stay, out in the middle of nowhere. Some of these dorms were sixty or seventy years old. These kids thought that this was being done because they were black. They rebelled and had a "sit-in" in the union. Someone said during the "sit-in" that the state police were coming with dogs.

What happened next?

Coble: Well, the kids rioted because they thought they were going to be attacked. For four years, they had hanging over them the fact that they could be prosecuted at any time because of this. It had been a misunderstanding on the kids' part, and a poor preparation on the university's part. When you bring in 200–600 black students in one year and you don't prepare them for college life and have them on a campus that isn't ready for them, you're asking for all sorts of problems. It made for an interesting situation.

So this was the beginning of your active role of assisting black students through the educational process.

Coble: I spent a year working to increase the number of black graduate students, but as black graduate students we told the administration that we would not bring in new black graduate students under the same conditions in which the undergraduates had come. I ended up spending a whole summer finding housing for 200 graduate students instead of doing my research.

It sounds like your commitment paid off for the University of Illinois. You also seem to be saying that race and gender were not a major issue for you when you were a graduate student at the University of Illinois. Please comment on this aspect more specifically.

Coble: When I was choosing a graduate school, I called Herman Branson, who was the chairman of the department at Howard. He said, "You know we [blacks] have had a lot of problems there. We have had students go there and have all kinds of racial problems." As we discussed it further, I realized that some of the racial problems were related to the social side of things. He was referring to the late 1950s, and some of the black males there dated white women. During that time, you just did not openly date white women, so a couple of people had some problems. My godfather had also gone there during the 1940s and had gotten his degree. So I had heard about these things and I knew basically what I was getting into.

Biophysics was taught in the physiology department at that time. The biophysics program was very welcoming. The department had between 50 and 70 graduate students at that time. My first interview was with the chairman of the physiology department, C. Ladd Prossor. Floyd Dunn was my thesis advisor. I remember there being some concern about my transition back to student life. They said, "You've been out of school for four years, so don't take an overload. You've been teaching and you are not accustomed to taking somebody else's orders in the classroom. That may be a problem." I got a lot of advice, but I wasn't worried. I didn't think being out of school for four years would make a big difference. I was fine.

When you graduated from the University of Illinois, did you go directly to Howard University?

Coble: Not exactly. I did go to Howard before I had finished with my dissertation, but I had spent two years at Washington University in St. Louis doing essentially a postdoc with Barry Commoner in the environmental science field. During part of the year, you could get frogs to do the research. We were interested in a certain species of frogs. We could get the frogs from

Texas at that time; because of the DDT problem in waters fewer frogs were available for the research project that we were trying to do.

What were you doing with the frogs?

Coble: I was looking at the effects of high-intensity ultrasound on the frog's skin.

Really? What did you find?

Coble: I found that one of the effects is that it separates the membrane potential measurement from the sodium transport measurement. There are two very different effects and apparently the membrane potential depends on the integrity of the whole system. Whereas the short circuit currents really just depend on what the sodium transport system is doing, so you can affect one without affecting the other.

Did you continue this research at Howard University?

Coble: No.

What was the impact of race and gender on your science career at Howard University? Were you the first black female in the physics department? What was the general makeup of your department?

Coble: Yes, I was the first black female. I came to Howard the year after Herman Branson left. We had only one female teacher other than me, and she was not black. She stayed only one year. The department makeup was mostly white when I first came back. Now I guess you would describe it as a more international mix of people. We are constantly trying to get more black faculty.

What was it like when the department was mostly white? Were there more racial issues?

Coble: Oh, yes. The gender issues were a little less important. Faculty issues were very much along racial lines. But it was not always race; sometimes, I think that the faculty was just a more conservative group of people. If you were not in a standard field of physics, say theoretical physics, decisions for tenure and promotion became an issue.

So how did that affect your tenure and promotion in the department?

Coble: I guess there were seven women in different departments who were given tenure, but we weren't given promotion.

You are tenured but you are not a full professor?

Coble: Yes.

You are tenured but you're an associate professor?

Coble: An assistant professor—and part of the reason is that I just haven't bothered.

How were they able to get away with that?

Coble: In the past the department initiated the tenure and promotion process. Now the faculty member has to. I just haven't done it.

You don't care?

Coble: Not very much. I've got five more years to teach and I can retire if I want to.

Okay, I see your point. You are viewing it from a different perspective than I. You have control over that choice.

Coble: Right.

Having some of the greats at Howard, like Mordecai Johnson and Herman Branson, must have made a difference

Coble: Yes, but Branson had left by the time I came back to Howard. He left in 1969. I returned in 1971. I was a student during Johnson's administration. Some of the faculty were Warren Henry, Halson Eaglesor, and A. Thorpe. They were helpful.

Beyond the race and gender issues that we have discussed, how was it to return to your alma mater?

Coble: I had a job back at North Carolina A & T University after I finished my Ph.D., but I wasn't sure if I wanted to be back in Greensboro. The people there were very nice to me, but I was in the School of Engineering and my interests were in the Liberal Arts. I went to Washington for a meeting and I decided to go by Howard University for a visit. The chairman and I discussed my availability. The officials at Howard University agreed to give me six months to complete everything. That was the agreement that was made. As I look back on that decision, I would not recommend anyone take a position without the degree in hand and some postdoctoral experience. Those are two advantages you should have when applying for an academic position.

Yes, that is true. It is very difficult now to get a job without that postdoctoral experience. But you actually had the experience from working in the environmental arena with Barry Commoner, right?

Coble: Yes, I did, but I also recognized that the demands of a new job are not always conducive to finishing a degree, especially the doctorate.

Did you feel you were home when you went back to Howard?

Coble: I went back to Howard because I thought I knew it. I knew Howard from the days of Mordecai Johnson and Herman Branson. It was not the same once I got back. James E. Cheek was now the president, and we were also under a Republican administration when I got there. It wasn't bad at first, but it got worse over time. There was a 30–40 percent decrease in federal money for grants to do research.

You mean millions of dollars were no longer available.

Coble: Yes, I am talking about million of dollars.

I thought that being at Howard University in the nation's capital would have been an advantage.

Coble: Being at Howard was an advantage, but you had to work with someone who already had a laboratory set up. When I arrived at Howard, the research that I was interested in was a more applied approach in physics. The applied and environmental area was not necessarily in line with the current departmental programs, which seemed to be more theoretically based. Later on, faculty were given their own start-up money.

It sounds like a mentoring component was lacking in the department. I think some departments do a much better job these days of pairing up junior faculty with effective, more experienced faculty members. Despite the lack of mentoring for you early on, you have been quite successful in doing so for your students.

Coble: I felt that I could do more for black students at Howard than I could at another place, where you may only have ten black students. I wanted to reach a mass of black students. Black students would be so isolated where the numbers are so few. It would have been nice for me to be there, but the number of lives I would be affecting would be minimal at a majority institution. At Howard, my impact would be experience by a larger number of black students.

Do you think the number of black women has increased in the physics department since you've been at Howard?

Coble: You have to realize that at Howard you are graduating fewer than six students with physics degrees in a year. That's considered a big class. We start out with 15–30 students who say they are interested in being a physics major. In the final analysis, maybe 8–10 students actually become majors, because many of them don't have the prior preparation in math to be successful in physics. A few of the students are prepared to do physics before college. We have students who are Phi Beta Kappa.

Most of the students who stay are very good; if they were not they wouldn't survive. We also get students who are not disciplined and they don't spend the necessary time to do the work. It takes a lot of time to do physics and unless you have a great background and skills, you usually cannot skip that part. Of course, there are exceptions to the rule. There are a few kids who have a great deal of talent and skill and don't have to spend as much time.

Poor preparation is still a social problem.

Coble: Yes, it is going to continue to be a problem. I will have a potential student come in and ask about being a physics major. I will ask, "What math have you taken in high school?" And they will say, "I've taken algebra I and business math I and II." So I will say, "Why haven't you taken geometry, trigonometry, or precalculus?" They will say, "My teacher told me I didn't need it because I am going to become interested in something else later." I will say, "Didn't you tell your teacher you were interested in physics?" And they say, "Yeah." That's happening all the time. This is what makes me angriest about the system. College professors believe these people who are career counselors, or even some high-school teachers, don't have these students' best interests at heart. On the other side, when you have 1,200 students and only two counselors, you can't expect so much interest. And many times the counselors aren't black, so there's no relating to the issues.

> *Those are very important points about how black students in particular are counseled early on. They are basically tracked out of science before they even start. There are some summer precollege programs, but sometimes they are only interested in the "cream of the crop" type students.*

Coble: That's usually why I get so discouraged about these programs—they are targeted to the students already at the top. I'll tell you a story. I had a classmate named Paul Brown. He went to Dunbar High School in Washington, DC, which was not the greatest school (in my opinion). He didn't get any encouragement because he didn't develop until later. He was a late bloomer. He came to Howard and had a 3.9 G.P.A. as a physics major. At Howard, he really blossomed and did well. We miss a lot of kids like that, who have a lot of potential, but no one nurtures that undeveloped talent. We don't reward black males. We actually reward black females better than black males for coming to class. Society protects the girls when they are progressing socially. Boys want to go play basketball. We still are not socializing the boys to understand that education and academic achievement are a part of maturity.

That's true. Have you seen an increase in the number of black women to complete the physics program at any level?

Coble: Yes, there has been an increase but not a huge one. We have had only three black females to complete a Ph.D. here. They are Elvira Shaw, Arlene Maclin and Helen Major.

What professional organizations do you participate in? Do you think they offer a service that you need as an African American female?

Coble: Well, I am a "sometime" member of the American Association of Physics Teachers (AAPT). The summer meeting is in the middle of the summer school session and it is hard to find someone to come and teach a whole week of your courses. A week off for a conference is a high percentage of the summer semester. I serve on committees when I can for my national organizations. I've been a steady member of the American Association of Science (AAAS). I have been associated with the Minority Women in Science (MWIS) since 1978. AAAS serves me better than my physics associations because my interests are much broader than physics. The AAPT has minority committees and reports. They have developed a video that is targeted to urban schools to introduce blacks to the physical sciences. The AAAS Black Church Project, which provides hands-on life science activities to children in the Washington area, has been a good program, but it doesn't reach all of our youngsters.

You mentioned the research that you had done earlier at Washington University. Have you conducted any other research?

Coble: I really haven't done that many publications. It wasn't difficult but it wasn't my major interest. I spent a year at the National Institutes of Health (NIH) looking at paracellular pathways. That research was presented at a symposium and published in a proceedings.

How has marriage or children impacted your career?

Coble: I didn't get married until I was 49, so marriage was not an issue. I guess one of the disadvantages of being in a field that is mostly male-dominated is that men don't always look for a wife in the same discipline. Occasionally, you will find someone who marries in the same discipline. On the whole, black male physicists don't want the competition of a wife in the same field.

By the time you were married, you had achieved quite a bit and were established. Was there any particular reason that you waited to get married?

Coble: As far as marriage was concerned, I wasn't really looking for a husband. Marriage just came by. I think that black women have a variety of experiences with spouses and family life while trying to balance a science career. We have female graduate students who come with children and that has an impact on them. We also have had students who are married, and their experiences are not different from those who are single—provided they have a support system. For women, the support system is crucial. Sometimes women do not have supportive husbands.

What advice do you offer young women who are trying to balance a science career and family? How do they survive the rigors of science?

Coble: They must have the strength of their conviction. The field must be interesting to them and students must use their strengths to be successful. Students can't allow anybody else to evaluate who and what they are. I have female students say to me, "My boyfriend doesn't like the fact that I am in the sciences." I ask, "Are you planning to marry him?" I explain to them that they must understand the limitations this kind of person might place on them in the future. You don't want anyone to limit your options and you shouldn't allow anyone to do that. Some young men feel intimidated that the woman is in physics and they also feel that they have to compete with her. Unfortunately, in the black community we are operating in an old or African system. That is, many of the young women allow these young men to determine who they are and define what their role will be. There's regret later among many of our females after they have become committed to some of these guys. They got caught up in his definitions or dreams before they could realize their own.

You express the issue very well. I am sure young women will appreciate your candor and sincerity about building a future in science or any career. What are your final comments?

Coble: We have to encourage each other to keep going, both young and old. We have to be willing to mentor. We must also not forget those who came before us. We have tremendous strength and history.

Interview Date: July 1996.

FREDDIE M. DIXON

One of Our Own

FREDDIE M. DIXON IS A PRODUCT OF TWO historically black universities—and proud of it. Like so many young African Americans who grew up during the era of segregation in the Deep South, she was influenced by the civil rights movement to become a success in life. She grew up in Jacksonville, Florida, in a middle-class, well-educated African American family, where she was strongly encouraged to pursue higher education. Freddie recalls, "My grandfather believed in educating his children and grandchildren. My family expected me to become a medical doctor." After high school, she went to Florida Agricultural and Mechanical University for a bachelor's degree in pre-medicine. However, Freddie

Freddie M. Dixon is chairperson of the Department of Biology and Environmental Sciences at the University of the District of Columbia. Her experiences demonstrate the power of the historically black college's role in developing America's science and technology workforce.

did not continue in the medical school track but instead received her doctorate in science. Although Freddie's family would have been happy if she had become a doctor, she chose instead to go to Howard University for graduate studies.

Freddie's story differs somewhat from that of many black women scientists. She did not experience many of the usual racial prejudices and common gender biases in her educational pursuits. Freddie believes that this experience can be attributed to the positive role models, mentoring, and strong supportive systems so commonly found at the HBCUs for African Americans—and really all students, regardless of race and gender. She thrived at Howard and received her doctorate in zoology with a specialty in cytogenetics. While at Howard, Freddie also met her husband, who has been supportive of her career.

After Howard, Freddie's journey took her across town to pursue a professional career at a technical institute in the District of Columbia. Later, this institute was merged with another local college to become what we now know as the University of the District of Columbia. Although Freddie liked her new job, her dreams of research had to be put on hold because of the heavy teaching loads. Many of her professional days were spent primarily as an instructor in the biological sciences. Though it took a while, she has been able to carve out a successful research program by finding supportive colleagues and collaborators within her own institution and at other institutions. For example, she became a visiting professor at Georgetown University Medical School in the Department of Microbiology and Immunology. In addition to this approach, she has been willing to explore new research areas. Her flexible approach has paid off. She has been the recipient of several research grants and publishes in some of the leading journals in her field.

Like so many other black women scientists, Freddie has worked tirelessly to train, educate, and motivate young people into productive careers. She has served as chair and president of several organizations that target young women and African Americans in the D.C. area for careers in science and technology. She has chaired the National Minority Women in Science network under the auspices of the American Association for the Advancement of Science. Freddie has never failed to give back to her community and to the university. She says, "Although I have worked long and hard to finally develop a strong research program, teaching is still a passion of mine." In addition to her teaching responsibilities, she makes sure her students at UDC participate in regional and national conferences so that they receive exposure to science and research early on because she understands the value of being mentored and supported. Although she still wears many hats in her scientific career, she is now doing what she dreamed of and continuing with her passion for teaching. Freddie Dixon is one of our "homegrown" products. Although some people may question the role of many struggling black colleges and universities in today's society, Freddie's story reminds us why the doors of black colleges and universities should remain open.

How did you first become interested in science?

Dixon: I am not sure. I think I had a natural curiosity as a child. I did well in school and my parents encouraged me.

So did your parents influence you to pursue science?

Dixon: Yes, my parents definitely encouraged me. They actually expected me to become a medical doctor. They knew that I liked the sciences. I grew up in a family that believed in education for the children. My grandfather made sure his children were educated. My mom and most of her siblings were college-educated.

I was a pre-med major but I was hooked on a science career when I took my first genetics course in college. I was totally fascinated by the subject.

What role did your teachers play in your selecting your career? Where did you grow up? Did geographical location have anything to do with your selection of or interest in science?

Dixon: There was no particular teacher during my K–12 years that led to my thinking about science as a career. I grew up in the south in Jacksonville, Florida. Of course, the south was segregated at that time, and that may have influenced my decision to do my graduate work at Howard University.

Where did you go to for your undergraduate degree?

Dixon: I went to FAMU (Florida Agricultural and Mechanical University) for my bachelor's degree. My degree was in pre-med, which was actually a major at the time. I chose it because the curriculum was more difficult than the traditional biology major. From there, I went to Howard University and received my master's degree in zoology in 1970. I continued in the Ph.D. program with a specialty in cytogenetics and I completed that program in 1973.

Do you feel race or gender played a role in how you were viewed as a student?

Dixon: No, I don't think it was an issue as a student. I had a mentor, David
Ray, who really nurtured me for five years during my graduate training.

> Do you think the fact that Howard is a predominantly black university
> had anything to do with your positive experience?

Dixon: Yes, I think that is accurate. The major problem that I had at that
time was being away from home. I was only twenty-one years old when I entered Howard University as a graduate student. Fortunately, I met David
Ray, and he really showed me how to maneuver my way around the university. At that time, there were so many prominent African American scientists there.

> What about issues of race and gender during your professional life?

Dixon: Yes, there have been issues, but I had an unusual beginning for my
career. Unlike the typical black female scientist, I actually began with a black
female as my supervisor. My department chairperson was a black female. I
had her along with other black women as role models. I would say that my
only issue was being new and young. I took some courses that I probably
shouldn't have taken, but that's part of being the new kid on the block. When
I first began, I was at a technical institute which was primarily a teaching college. We merged to become the University of D.C. in 1977.

Earlier, when I applied for jobs, I think I did not receive some of them
because I was black. In recent years, I have observed how other scientists
have treated me in certain scientific settings. For instance, I was at a meeting
in the Midwest and I found that my white male colleagues were very distant
to me and another black female colleague, but after we presented our work
their attitude seemed to change.

> Do you find it difficult to distinguish between race and gender issues?

Dixon: I think it can be difficult. I had a black female colleague who was
propositioned by a colleague at a national scientific meeting, and we really
don't think that would have been the case had she been of another race. I
think that is because they don't see us as scientists. Sexual harassment is
something that we tend not to talk about or deal with in the scientific world.
It does happen. I think that would be more gender-related when any female
is harassed. However, I recently dealt with a situation where a young female
student referred to me as a girl. This happened at a predominantly white institution where I was a visiting professor. I was very taken aback with that
kind of behavior. After all, I am a seasoned scientist and I have been doing

my research there for quite a while, so it was a bit surprising to still see this kind of behavior. That may have been because I was both black and female. Of course, I had a long talk with her about the incident and she apologized. It is generally difficult to distinguish the reasons why you see certain behavior directed at you and not at others. Except for those incidents, I have had a great working environment.

Are you the first woman to chair your department?

Dixon: No, there have been four other women before me. I have a great relationship with my colleagues. As a faculty member, I always worked well with others, so now that respect is returned to me as chair.

Do you feel the civil rights movement played a role in your success or in your selection of science as a career?

Dixon: The movement played a role in my success, but not in my selection of science as a career.

What about the women's movement; do you think it played a role?

Dixon: No, I don't think so.

How do you think other scientists can be more supportive of black women scientists?

Dixon: I have been very fortunate in receiving support. Earlier I felt that men should be more supportive of women in science since they are mostly in the power roles. I have since changed my thinking on that. I think it is a matter of getting with the right person (male or female) who is supportive of you and your research. I have a white male colleague who is very supportive of my current research program. I started out working with Inez Bacon and she was quite helpful.

I think it is also a matter of being self-motivated. I mostly taught when I first began my career. I wanted to do research and I applied for several grants and was turned down but I never gave up. I was encouraged by Dr. Preer to write proposals and get them funded through our agricultural experiment station. He was one of the directors of the research at the university. I even wrote a proposal two times before it was finally accepted. I have been able to finally carve out a research program for myself, but it has taken persistence and hard work. So part of the responsibility lies with the black woman scientist.

How have marriage and family played a role in your life as a scientist?

Dixon: My kids come first and the marriage is second! I am joking, of course. My husband has been very supportive of me throughout my entire career. I have never let my professional career interfere with my family because they deserve me more than my career. I feel strongly about being there for my children, even though they are grown now.

> *Are there any particular strategies that you would recommend for young women who are trying to balance a science career and family?*

Dixon: It is great when you have a supportive spouse. I had my son before I finished my Ph.D. My husband did a great deal of the care for him, like taking him to the babysitter. I did not have the worries that many working mothers might have because I had such a supportive spouse. I would say that would be the key, along with good daycare when the children are young.

> *What can we do as black women to change our image in science? How do we become more visible?*

Dixon: We need to participate in our scientific societies and network with other scientists who are willing to work with us. We need to form collaborations to get our research done. We need to be active in our own groups, like the minority women in science network. We need to do our part and make sure we mentor young people.

I have done a great deal of this work over the years. I also try to become visible by attending scientific meetings. I go to the northeast Agronomy Society meetings almost every year and I try to present a paper at that meeting. It can be a national or regional meeting, but we must be active participants. I also attend the annual meetings of the National Institute of Science, Beta Kappa Chi Scientific Honor Society, and the Brookhaven Semester Program, where undergraduate students present papers. Usually a student from my laboratory presents our work.

> *How do you envision your future as a scientist? Where do you see yourself in ten years?*

Dixon: I don't know. I hope I will still be doing research. I think research will be my main focus. I always mentor undergraduate research students. Teaching is also my passion.

> *What do you think your greatest contribution to science will be?*

Dixon: Students have been a great joy. Most of the students come from the Washington, D.C. area. A lot of them have problems because of family obligations and they need to work. It is a sacrifice for them to come to school. Most of them are poor. They face just about any situation that you can imagine. Some of them use their lunch time just to take my class. Many have the goal of getting their B.S. degree and going to graduate or professional school. So I know with all their issues I have to make sure they are really prepared for what is ahead of them. They know that when they come to my class they will have to work and get to know each other. I feel that there is no greater accomplishment than when they have become successful and they come back to tell me. It is a beautiful feeling when you have been a part of shaping someone's life in that way. You know where they came from and how far they have come.

What advice do you offer to young women, especially young black women, who are interested in science careers?

Dixon: I would tell them to remain poor and finish their education. I would tell anyone to go beyond the B.S. degree in biology or any science field. I would not recommend getting a professional job before one's education is completed. I have seen it so many times when former students get a taste of the money from the first professional job and never return to school. This is something that they usually regret. That is one piece of practical advice. I would also tell them to believe in themselves, network, and work hard. I've been successful because I have been self-motivated. I also have been in very nurturing environments, so I would advise trying to find a supportive environment to pursue a science career.

Interview Date: August 1996.

Selected Publications and Research Activities

Cousin, C., J. Grant, F. Dixon, D. Beyene, and P. Van Berkum. 2002. Influences of biosolids compost on the bradyrhizobial genotypes recovered from cowpea and soybean nodules. Archives of Microbiology, 177(5): 427–430.

Sturtevant, J., F. Dixon, E. Wadsworth, J. P. Latge, X. Zhao, and R. Calderone. 1999. Identification and cloning of GCAI, a gene which encodes a cell surface glucoamylase from Candida albicans. Journal of Medical Mycology, 37: 337–343.

Dixon, F., M. R. Calderone, and J. Sturtevant. 1998. Isolation of a putative glucoamylase gene (CaGAM1) from the human pathogen Candida albricans. American Society for Microbiology Annual Meeting, 115.

Dixon, F., J. R. Preer, and A. N. Abdi. 1995. Metal levels in garden vegetables raised on sludge compost amended soil. Compost Science and Utilization, 3: 55–63.

Dixon, F., A. G. Fall, and W. O. Oguya. 1992. Metal concentration of soil and tree foliar in forest stand treated with sewage sludge. Northeast Branch ASA, Storrs, CT, p. 18.

Dixon, F., J. Preer, and A. Abdi. 1991. Metal levels in vegetables grown in soils treated with sludge compost. UDC, AES Bulletin, 7: 1–10.

Selected Research Funding

National Science Foundation, SMET Research and Training Center, Co-investigator, 2001–2002.

United States Department of Agriculture, Agricultural Experiment Station, Integrate Pests Management, Project Director, 1998–2001.

NASA, Anacostia River Institute for Remote Sensing, Training High School Students, Co-investigator, 1997–1999.

United States Department of Agriculture, DC Agricultural Experiment Station, Effects of Rhizobium on Crop Yield in an Urban Composted Biosolids Amended Soil, Project Director, 1995–1998.

Department of Health and Human Services, Public Health Service, RIMI Molecular Biology Center, Assistant Project Director, 1995.

United States Department of Agriculture, Utilization of Sludge Compost on Forest and Non-Agricultural Land, Principal Investigator, 1987–1992.

ELVIRA DOMAN

A Class Act

ELVIRA DOMAN CARRIES HERSELF WITH STYLE, grace, and dignity. A graduate of New York City's Ophelia DeVore Charm School, Elvira said growing up in New York during the 1940s made a big difference in her life. Not only was she able to receive one of the best educations during that time but she also had the opportunity to take advantage of the charm school and the public library. She credits her mother with exposing her and her siblings to all that the city had to offer, including her own library card by the age of five. Dancing and voice lessons were also a part of her early training. In addition to her mother's positive influence, community people, like her Sunday School teacher, played an important role in her development.

Elvira Doman, retired program director of the Integrative Animal Biology Program at the National Science Foundation, is a scientist who combines determination with grace and dignity.

An outstanding student in high school, Elvira decided to continue her education at Hunter College, where she majored in chemistry. Always up to any challenge, Elvira recalls, "I decided to major in chemistry because I thought it was more challenging than biology." She excelled in her studies and pursued other interests as well. She had to learn early on how to balance a challenging science major with church responsibilities and part-time jobs. This skill proved to be helpful in her dual career as scientist and administrator.

After graduating from Hunter College, Elvira received two master's degrees; one in biology from New York University, and one in biochemistry from Columbia University, breaking down some barriers along the way. She recalls becoming weary of being told constantly that she was the first black graduate student in biochemistry. Always a talented and gifted student, she did not escape the challenges of being female and black—and the effect of the glass ceiling. She recounts being cheated out of authorship for publication and jobs that she is certain that she was more than qualified to handle. Through all of this, she did not give up; she defied the odds and received a Ph.D. in physiology and biochemistry from Rutgers University in 1965. Along the way, she married and had two lovely children.

Elvira began her career as an academician at Rutgers and Seton Hall Universities. In 1978, she had the opportunity to change career directions when she was offered a position as an assistant program director at the National Science Foundation. She accepted the position and progressed through the ranks to become program director of the Integrative Animal Biology Program. She faithfully served NSF for twenty-one years until her retirement in 1999. Always concerned for others and a mentor to many young scientists, she received high recognition of service from Cornell University in 1999 and several honors from the National Science Foundation, including the Director's Equal Opportunity Achievement Award. In fact, she was one of the few African American women to ever serve as a full program director for NSF. She has mentored students in the famous Meyerhoff Scholarship Program at the University of Maryland, Baltimore County. In addition, she has served as the president of the D.C. Metropolitan Organization of Black Scientists—just to mention a few of her community accomplishments.

An avid reader and accomplished golfer, Elvira relaxes and enjoys life with her friends and family. She says it is always a joy to see granddaughter Serena. Still a woman of style and elegance, she has appeared in a fashion show for senior citizens of Washington, D.C., many times. When she is not busy with those activities, you can find her raising her voice at Shiloh Baptist Church. A class act in her professional and personal life, Elvira is truly a woman of dignity and grace.

How did you first become interested in science?

Doman: According to my mother, I became interested in science at the age of twelve when I was probably in the sixth or seventh grade. I came downstairs one morning and told her that God had spoken to me in a vision and told me to become a scientist. This incident is memorable in my mind because there weren't any scientists in the family or the community that I can recall. (I always hesitate to answer this question because the answer has a religious origin which some may not understand.)

That's an interesting response.

Doman: I was one of those people who view difficulties as challenges. Mathematics was supposed to be hard for females. Thus I was attracted to it. Not only did I take math, but I also did well in it. When my math teacher learned that I was completing all of the homework with facility, she assigned additional problems at the back of the chapters for me to do at home. She wanted to keep the kid busy.

I had a good mother. My mother took us to the public library very early to join—somewhere around the age of five. We all had our own library cards at very early ages. I skipped kindergarten because I knew how to read when I got to school.

How many siblings did you have?

Doman: There was a total of four. I have two sisters, and my brother is the eldest of us all. I am the only one of the four who attended college and took advantage of the educational opportunities that growing up in New York City afforded.

Was there any particular family member who encouraged you in science?

Doman: Yes, my mother was very bright and supportive, even though she was neither college-educated nor a scientist. She was an attentive stay-at-home mom, so I was very blessed.

I recall her sending us to the nearby Union Settlement House, where we were given lessons in voice, folk dancing, interpretative dancing, and ballet for a small cost. The settlement houses of those days are similar to the community centers that exist today in many neighborhoods. I took piano lessons at an early age and excelled as a student so that I was given an opportunity to play at Carnegie Hall. My mother was supportive of all four of us in whatever we chose to become. My father was passive about my career decisions. I am still grateful that neither he nor my mother ever tried to prevent my selecting science as a career. I learned later in my life that a lot of my peers had family members who said to them that majoring in science was very difficult and they should not take it because it required too much in the way of brainpower. This type of comment from family members that is heard early on in life is what discourages many of our young people. I'm glad that I did not hear such negative comments from my parents.

I heard a guest speaker at my elementary school—I don't recall her name, but I remember wanting to attend Hunter College because of her. I was so impressed with her presentation and the way she conducted herself that I wanted to attend that college.

> *That's the power of role models. Was there anyone else in the community?*

Doman: I had a supportive Sunday School teacher named Bertram Glass. I grew up in New York City at a time when black history was not celebrated or made available in the public schools. You learned about George Washington Carver and that was about it for blacks in science. Fortunately for me, Mr. Glass provided his students with black history lessons on Sunday mornings based on articles published in *The Amsterdam News, The Pittsburgh Courier,* and black magazines such as *The World.* I don't recall whether or not *Ebony* magazine was included. Mr. Glass brought the achievements of blacks to our attention. He was an ordinary Sunday School teacher with extraordinary vision for his youth. I bring him up because I consider him a part of my family. I would say that he served as a surrogate father in some ways. He taught me how to drive an automobile and how to play tennis. He introduced me to a world I would not have known otherwise. My own father did not complete high school. He was busy working at night and slept during the day. So Mr. Glass, my Sunday School teacher, played an important role in my life.

> *What about teachers in your K–12 experience? What role did teachers play in your selection of science as a career?*

Doman: Hunter College High School, commonly referred to as Hunter High, was one of the top schools in New York City, with very high academic standards. You had to take an admissions test which included English and mathematics. In addition, you had to obtain approval from the faculty members of the junior high school. I recall (unhappily) that my English teacher refused to sign for me because she felt that if I took the test and failed, it would place a black mark on the school. Fortunately, another teacher stepped in and saved the day for me by signing for both subjects. I took the tests and passed with flying colors. I was the only one of four students who qualified. Consequently, that teacher's role was pivotal in my being able to pursue science. By attending Hunter High, I received the best public education that a poor person from the slums of New York City could obtain. I attended school with girls from wealthy families and girls from all five boroughs of New York City. We were all about at the same level of intelligence and academic achievement. By attending that preparatory school, I was as confident as anyone could be that I could achieve anything when I graduated.

Did geographical location play a role in your selection of science as a career?

Doman: Yes, as I reflect back I find that there were many advantages in growing up in New York City. I had a very good public school education and a good selection of affordable yet excellent colleges and universities. All that was required was a good academic standing. Academic standing was based on letter grades in those days. One had to have an average of B+ or A to be considered academically qualified for admission to the four city colleges. I would not have had the golden opportunity of a free college education had I been from somewhere else. A free college education was critical to my success, since my parents could not have afforded to put me through college. There were also many other opportunities in New York City that helped build my life and career. For instance, I made the Honor Society both in junior high school and in high school. The recognition of my academic achievement helped build my self-esteem and confidence to pursue mathematics and chemistry.

I was also active in sports. I was elected president of the Athletic Association at Hunter College High School, and participation in sports activities was important in terms of developing confidence and team-building work skills. I also had the opportunity to attend the charm school that was headed by Ophelia DeVore. I doubt that I would have had this experience outside of

New York City at that time. Learning how to present oneself properly both mentally and physically to the public has always been important for people of color. And the training I received at Mount Olivet Baptist Church in New York City helped me to learn how to be a leader. I became a Sunday School teacher at the age of seventeen. This helped me to give something back to younger people in my church and community.

When did you begin college?

Doman: I attended college in the 1950s. I was accepted at Columbia University, but there was no money to go there. Hunter College was a public institution, so I could attend with free tuition. I went to Hunter and majored in chemistry because I thought it was more challenging than biology. Most women at that time majored in biology. Of the eight hundred women in my graduating class, only eight of us majored in chemistry and I was the only African American.

What was one of the most important skills you learned while attending college?

Doman: There were many skills that I developed during my college years, but I think there is one that stands foremost in my mind because it was essential to my success in graduate school and then my professional career. That skill was balancing numerous interests. I believe it's called multitasking today. As a chemistry major, my study schedule was demanding. My studies, for example, included a three-hour afternoon session twice a week. I was a cashier-wrapper/salesgirl at Macy's three times a week on a part-time basis. I participated in church choir recitals and parades. I had to learn how to get the work done well and utilize my time effectively and efficiently.

Do you feel that race and gender played a role in your selecting science as a career or that they have influenced your career?

Doman: Growing up, the first time I encountered "color" or race problems was when I attended elementary school. I graduated at the top of the class and was due to receive an award from the Daughters of the American Revolution. But I recall that the woman who was supposed to present it to me did not attend the ceremonies for whatever reason. Another incident occurred while I was in junior high school. Although I competed and won first place, I was not given the art award for reasons that elude me today. I was instead given a "compensatory" gift with the excuse that the other girl needed the award in order to get into an art high school. They felt that after all, I had won all of the other academic awards.

Another eye-opening experience for me was when I applied for a position as a junior chemist at a pharmaceutical company in upstate New York. Based on my achievements, I was telephoned and invited for an interview. The application made no reference to race. When I arrived, the interviewer was shocked to see me and headed into a back room closed off by a glass wall to make a telephone call. I overheard him say, "I'll get rid of her in a minute." He then proceeded to give me all kinds of lame reasons for not hiring me. I had decent grades and was even on the dean's list at Hunter College. Later when I attended Columbia University College of Physicians and Surgeons, I grew weary of being told that I was the first black graduate student in biochemistry.

Maybe the obsession with my race wasn't racism, but it certainly was annoying when the same references were not made to other colleagues. Early in my career, I worked in a laboratory at Sloan-Kettering Institute for Cancer Research, and one day I discovered that my name was not listed on a publication but the other technician, who was white, was listed. I had performed all of the calculations, laboratory work, etc. My supervisor's response to me was, "You don't have an advanced degree." The other technician did not have an advanced degree, either. My response to my supervisor was, "The next time you see me you will have to call me 'doctor.'" Four years later, I returned to his office with proof of a Ph.D. in physiology and biochemistry.

I have encountered so many "isms" along the way. When I applied to the Ph.D. program, the interviewer asked, "Why are you applying? Don't you get enough attention at home?" That was more gender-related than race.

How did family play a role in your science career?

Doman: I was married early on and had children. As for any working mother, a good support system is necessary.

What type of support system did you have?

Doman: I received support from all kinds of people. I've already mentioned my mother, who was the greatest influence in my obtaining a good education. The black church also played an important role. In addition, there is the mentoring and leadership training I received from Mr. Bertram Glass and other adults at the church of my youth.

I still recall participation of the beautiful black women in the community when I was in graduate school in New Brunswick, New Jersey. The women of the church provided support by having me visit their homes for Sunday dinners. One beautician even provided free hair care while another donated clothing. In addition, my sister, Lucille, who lived in Newark, allowed me to visit her so that I could take a break from classes. My niece, Linda, would let me sleep in her bed for an afternoon nap. She was so honored that her auntie from graduate school was spending time with her family. Then there were personal friends who were very supportive. When I ran into what I perceived as a racial problem on campus, they would say to me, "Don't worry about the grades or money. Just hang in there. You will be okay." They, too, had run into all of kinds of difficulties when they were in pursuit of higher education.

> *Are there any particular strategies that you would recommend when a woman is balancing a science career with other commitments?*

Doman: I would like to advise young women to get their education out of the way first. They should not give up an opportunity to go to school in deference to members of the opposite sex. In fact, these days men have more respect for a woman with an education than a woman without one. It's important to learn how to make sacrifices for priorities in life. A young woman can complete her schooling fairly quickly if she remains focused on that goal. For example, because of careful planning and good organization, I was able to complete the writing of my Ph.D. thesis in just six weeks.

> *So you recommend young women, especially black women, to get their education early and make themselves their number-one priority. You sound like my mother! What can black women do to become more visible in the profession?*

Doman: I have to admit that we are sometimes treated as if we were invisible. As a result, I cannot overemphasize the need to publish, publish, publish. Publications are very important for establishing a reputation as a knowledgeable and productive scientist, not only in the workplace but throughout the scientific community. A list of publications serves as a basis for funding as well as for employment. If you have a good track record, you will be recognized in due time.

It takes about ten years of dedicated high-quality work before people begin to recognize your ability. Since written work is so important, I have followed the advice of my daughter, Paula, who is an attorney. Paula advised me, "Mother, make sure your writing is clear. Moreover, always put your best foot forward by proofreading all work-related documents several times, including informal correspondence such as e-mail messages." Visibility can be tremendously boosted by becoming a valuable employee. A valuable employee is able to carry out those assignments that others find too difficult or labor-intensive.

> *Speaking of volunteer activities, what role has membership in professional societies and organizations played in your career?*

Doman: I belong to several professional societies and organizations. They include the American Chemical Society (ACS), the Association for the Advancement of Science (AAAS), the New York Academy of Science (NYAS), etc. I also hold memberships in the Association of Women in Science (AWIS), Minority Women in Science (MWIS), and the D.C. Metropolitan Organization of Black Scientists (OBS). One favorite activity for me is the annual Science Discovery Day that is sponsored by the Minority Women in Science (MWIS). This event is held every year on the campus of either Howard University or of the University of the District of Columbia. We are reaching out to students at the junior high school level. This is a time when we can best reach them and help them prepare for a career in science while they have time to take the right prerequisite courses for college. There are also other outreach activities in which I have participated. They include judging science fairs, participating as a role model for science career day, or serving as a speaker. In addition to serving as the chief pre-health professions advisor at Seton Hall University for three years, I have mentored several students and young professionals over the years. My mentees are doing well. Of course, you know LaVern Whisenton-Davidson. I am very proud of her and her achievements.

> *Yes, I know LaVern. She is a sister in science for me. How do you envision your life as a scientist in ten years?*

Doman: Retired! I will not be sitting idly in a rocking chair. I plan to play as much tennis and golf as possible. I also have many church activities I wish to undertake. In addition, I will continue to help young people pursue a career in science.

> *Sounds like a wonderful plan. What is your greatest contribution to science?*

Doman: I am a person who cares about people. I firmly believe that there are a lot of innately talented and gifted people of color in the world. All they need is the same opportunity, guidance, and mentoring that is given to everyone else. With these tools, they can rise to their true potential. I have dedicated many years to going the extra mile, not only to seek out these "diamonds in the rough" but also to help them move to the next level. My support has ranged from offering words of encouragement to introducing them to potential mentors and including them in programs that offer visibility and growth opportunities. As a science administrator, I have tried to help make sure that talented, hardworking, productive scientists of all colors receive the funding and recognition that they truly deserved.

What is the final advice or impression that you want to leave with young aspiring women scientists, especially young black women?

Doman: Stay the course. Do not allow distractions, criticisms, and negative words of advice from other people to deter you from your goals. Remain in touch with your innermost selves. That was my secret in achieving the most important goals in my life. I was focused and determined. I ignored the advice that seemed counterproductive to my career choices. I can recall being told that Columbia University did not accept "coloreds" (that is what we were called in those days). My reaction was, "Let them tell me that to my face." I applied for admission and was accepted. I received a master's degree in biochemistry from that institution. Decide to stay the course—plan ahead, get good grades, and take advantages of golden opportunities.

I would like to add for those attempting to raise a family, they should bear in mind that there are sometimes opportunities for employment known as job-sharing. While employed as a part-time lecturer at Rutgers University, I was able to teach in the mornings while another colleague who was also part-time taught in the afternoons. The two of us shared the position with the title of lecturer. The university's daycare center in which I placed my son was in a nearby building.

It can be done. Success is ours if we are really determined.

Interview Date: August 1999.

GEORGIA DUNSTON

It's in My Genes

IF YOU EVER MEET GEORGIA DUNSTON, you will never forget her because her deep passion for life and love for her research permeate her very essence. Standing about six feet tall, she is a kind and gentle spirit, yet a strong black woman who speaks with clarity and vigor about her work on genetics. For almost thirty years, she has worked tirelessly to establish a leading laboratory in immunogenetics and the National Center for Genomic Research at Howard University. The laboratory that she and others have established at Howard University is the only accredited lab at a historically black college or university—and she is the laboratory's director.

Born in 1944 in Norfolk, Virginia, Georgia, like so many young black girls, grew up in a poor family that was strong and supportive. She credits her

Georgia Dunston, a professor in the Department of Microbiology at Howard University's School of Medicine, is the first black woman chairperson in microbiology at Howard University and researches the genetic code for her race.

junior high school teacher, Margaret Saunders, with recognizing and encouraging her in an annual math and science competition. With this early experience, she became oriented toward a career in science. Because Georgia did well in school, she was offered a full scholarship to Norfolk State University in 1961. It was no surprise to anyone when she majored in biology.

After graduating from Norfolk State University, Georgia thought she would get married and start a family, but life had other exciting plans for her, with a few twists and turns. During the summer of 1965, Georgia ventured to New York City to find her fate in life. Armed with her bachelor's degree, Georgia was excited about finding her way in the world and a job. To her dismay, she experienced discrimination due to her race and her gender. She recalls:

> The whole summer I looked for jobs. I signed with the employment agencies and I checked the papers. The paper was full of jobs but when I arrived for an interview the job was either filled or I would be directed to a maid or janitorial position.

This experience was very disappointing for a young, qualified woman ready to take on the world. Feeling dejected, Georgia returned home and visited one of her old professors at Norfolk State. That visit changed the course of events for her. Although the summer was practically over, her professor suggested that she enroll in the master's program at Tuskegee Institute (now Tuskegee University), where she could work in the prestigious Carver Research Laboratories. This idea was very appealing to Georgia given her current options. She jokingly says, "This is how it has been all my life. I never had a 'zillion' options for a career but I always had some kind professor or teacher who would tell me about that one opportunity to advance my education." And advance she did. Because Georgia was wise enough to accept the offer she had from Tuskegee, she embarked on the field of genetics.

During her stay at Tuskegee, Georgia met a visiting professor from the University of Michigan who taught her biochemistry. She made a good impression on him, and he encouraged her to consider the University of Michigan for her doctorate. Michigan was a different experience and presented a few challenges for Georgia, but it was nothing that she could not handle. In fact, she had an overall positive experience and still reminisces about her days in Michigan with former colleagues that she meets from time to time. She also made history at the University of Michigan by becoming its first African American to graduate with a Ph.D. in genetics in 1972.

With her well-deserved Ph.D. in hand, Georgia had some decisions to make—what to do and where to go with the rest of her life. In an honest and candid fashion she says, "I was tired. I had gone from first grade all the way to the Ph.D. Michigan had been a positive experience for me, but I had missed living in the South. Frankly, I missed my black community." She had been offered a postdoctoral fellowship in Salt Lake City, Utah, but it wasn't appealing. Again, fate would deal Georgia another, more interesting and challenging choice in life.

She was attending a national conference and taking a short break from meetings. As she sat to rest her tired feet, she glanced up and happened to notice some young black men looking in her direction. Sure enough, the young gentlemen introduced themselves and said that they were scientists from Howard University. This led not only to a beautiful conversation on the latest scientific discoveries but also to another career opportunity. Howard University was trying to build a program in microbiology and needed someone with expertise in genetics. Howard University was the ideal place for Georgia. It was the beginning of a long and outstanding career for her.

At Howard, Georgia began to answer some of the questions that she had pondered even as a child. She has done some fascinating work in immunogenetics, including some of the first studies that take into account race in organ transplantation and in the human genome. She has worked on different aspects of the human leukocyte antigen in blacks for over twenty years. She has been highly creative and undaunted in her pursuit of the highest possible quality research on the genetics of African Americans as part of the quest to discover and unravel the human genome.

In addition to her research, Georgia has taught, mentored, and advised hundreds of students through the Department of Microbiology and School of Medicine. She encourages young people to seek out mentors and study what intrigues them and captures their imagination. That's what Georgia Dunston did so many years ago back in Norfolk, Virginia. Her quest has been a lifelong search for answers to questions about the universe and the well-being of all people. One could say that Georgia's fascination with intriguing and probing questions is just in her genes.

How did you first become interested in science?

Dunston: It was from the influence of a science teacher, Margaret Saunders, in junior high school. We used to participate in a math-science conference and it was a big activity getting projects ready. The biology students had a little competition going with the math students. My earlier interests were oriented toward biology.

Did any of your family members influence your interest in science?

Dunston: No. At least, not in terms of science.

Where did you grow up? Do you think your geographical location influenced your selection of science?

Dunston: I grew up in Norfolk, Virginia. In retrospect, my geography did influence my selection of science, although at the time I didn't realize it. My hometown was segregated.

Because of the sensitivity to race, I had a curiosity about the differences in people. Why we were different, different in skin color, why our hair was different were the questions that I often pondered. I say that in retrospect, because I ended up in genetics, which is a science of variability. The kinds of questions that I posed and the advice that I got early on were really what directed the path that I took. I was interested in knowing why things were the way they were. It sounds trite, but I had a deep curiosity. I did fairly well academically in school and my science teachers were a big influence, but also my English teacher was important, too.

I was also in school during the age of Sputnik, and the space program was really taking off. I remember being questioned about career choices at that time. I was really more interested in philosophy than in science. I was interested in knowing how we know what we know, and how we know that that's the right thing. I guess it was part of trying to understand the differences of what's important and what's not. I enjoyed reading the classical philosophers and the questions they dealt with: who are we, why we are. Those things intrigued me. I remember my first counselor in terms of career choice and my teacher told me, "Those are interesting questions that are challenging and important, but you'll never get a job as a philosopher because you have to think in terms of where you're going to be able to get a job." The advice was to choose a science career because I'd be able to get a job there. I was a good student in science, and so this is the direction that I was encouraged to go. So I followed biology because I was interested in differences in people.

Where did you go to school?

Dunston: My bachelor's degree was in 1965 from Norfolk State University; my master's was in 1967 from Tuskegee Institute (now University), where I worked at the Carver Research Foundation. We called ourselves "the rat pack." We studied the DNA in pigeons. I received my Ph.D. from the University of Michigan in 1972 and I did my postdoctoral training at the National Cancer Institute here in Washington, D.C.

So you had teachers who were a positive influence on your selecting science as a career.

Dunston: Yes, I had teachers along the way who influenced me and recognized my potential. I mentioned Margaret Saunders, who had an early influence on my interest in science. Later a professor in an exchange program at Tuskegee encouraged me to pursue my doctorate. My exchange biochemistry professor said, "I think you're a good student and I think you should pursue your education." That's how it has been for me every step of the way from elementary school. I remember my elementary school teachers recognized my interest in my studies and encouraged me. My high school teacher recognized my potential. College really wasn't even a consideration for me because I wasn't in a family where higher education was a part of the norm. At that time, they were happy that I finished high school. I ranked in the top five in my class and I had a full scholarship to a state school. That's how college came into the picture.

Tell me about your undergraduate days.

Dunston: I went to college and I had a teacher who came from Ohio State who took an interest in me when I was a biology major. He was the first person who ever gave me a C, and I knew I wasn't doing that badly. He basically said, "You got a C because you can do better." This was kind of an awakening for me. He was the one who demanded more from me than just an acceptable level of academic performance. When I got my B.S. degree, I really thought, coming from the world that I came from, that getting a college education would open all the doors. I was going to go out and get the job I wanted and live the life I desired. So from Norfolk, this little lady got her degree and went to New York City to live with an aunt and take on the world.

What happened in New York City? Do you think race or gender played a role in the way you were viewed as a scientist?

Dunston: Let me go back a bit. After my undergraduate program was completed in biology, I decided to go to New York to take on the world, so to speak. That was a very eye-opening experience. That whole summer I looked for jobs. I signed on with employment agencies and I checked the paper. It was full of jobs. I wanted a job at the level of a laboratory technician position. I had a biology degree and they had the jobs, like hospital positions, laboratory jobs, etc. I went to the employment agencies and applied for many jobs which I felt qualified for. After a while, the behavior of the employment agencies started bothering me. I had grown up in Norfolk, Virginia, where the racial lines were clear and you kind of knew what to do and what not to do. I went to New York and I thought, that means I can get a laboratory job in a hospital. But I wasn't being offered anything that I needed any kind of education for. I'd go for a lab tech job interview and they'd tell me they had something in housekeeping or the job just closed. It was my first taste of a subtle kind of discrimination. I was told the job wasn't available, or at least the one I wanted wasn't, but they had something else. I wasn't prepared to take that, and I came home disillusioned. The whole summer I had looked for a job. I came back home and I went to visit my professor in August. I was dejected, and he told me, "I really think you should continue going on through school, and it's not too late. Tuskegee Institute (now University) is still advertising that they have Carver Research Fellowships." It was in a science journal. That was in August, and school was to start later. That was a long-about way of telling you how race played a negative role and how I found myself in graduate school.

What about other experiences concerning your race or gender?

Dunston: I think race plays a role. I think that above anything else, people have their preconceived notions before they get to know you. Those notions are based on generalities. If they don't know you as an individual, then you are treated like the group. I have to admit I was in Michigan because of race. I would accept race as a factor in being accepted into the graduate program. Tuskegee Institute had an exchange program with the University of Michigan. I had an exchange professor who taught me biochemistry at Tuskegee. I did well, and he asked me if I was going to continue with my education. At that time, I had no plans to continue because I was "in love." I was supposed to get married and live happily ever after with the white picket fence and everything. As life has its turns, that didn't happen. So, I decided to go to graduate school. Race was positive to my being selected to go to graduate school.

What was Tuskegee like? How did you end up at the University of Michigan?

Dunston: Well, this is another love story of sorts. I met this young man when I first arrived at Tuskegee and we were very close the whole time, and I just knew we would graduate and go to the next obvious step. But it wasn't to be. I didn't know what I was going to do. My biochemistry professor suggested I go for my Ph.D.; that's where Michigan came into the picture. It's like I never had thousands of options. There were these circumstances, and I had one option. Michigan would be the first time that I would be away from a Southern setting. After the New York experience, I was worried. This is the first time I'm away from family. I didn't stay long enough in New York for that to count. I'm sensitive and I thought I was carrying the whole weight of Black America on my shoulders. I was the first black in the department.

I really think the University of Michigan's faculty in my department had a discussion about the current situation in America. That is, the times dictated that they have at least one black in the program. I felt like they were saying, "We have to face the fact that we're going to have one." I felt like they drew straws and my advisor got the short straw. I mean that's how I felt! I felt like they were saying, "It's something we can't avoid. We may as well face the music, so let's see who is going to be stuck with her." My advisor was one of those who felt it was a waste of time to invest in women. He felt that investing in these women was never really productive because they just go off and get married. Because I was a black woman it was more of a question as to how my career was going to unfold. Anyway, we went through a transition but it ended positively. In retrospect, I've learned that it's a matter of difference in judgments based on perceptions, and after you get to know a person those initial perceptions can change for the better. I was told that because he was so elated that I finished I was the only student for whom he had a graduation affair at his home. I don't know if it's true, but Michigan was a good experience for me.

> *So it was a matter of gender and, perhaps, race at first. It seems that the two of you moved beyond the issues of race and gender. This was in the 1970s, right?*

Dunston: Yes, I think you are right in that assessment. There were lots of activities with the undergraduates and in some of the graduate areas for minorities and women. The social science areas had a lot more consciousness in that regard. The program I was in demanded and took my attention. I truly felt that my academic performance would reflect on blacks, so I felt on the spot to do really well. There was one course where I felt there was no way I was going to be successful. I was having difficulty on the exams. There were some dynamics going on that I didn't think were going to lead to anything, and I was really dejected and I had told myself, "Maybe they're right. I don't belong here. I can't do this. I can't keep up." It was the first truly academic challenge for me. I had been challenged, but I never thought I couldn't do it. I never thought it was beyond my capacity. I was dealing with that, and I went and talked to the chairman at the time. He was like a father figure and he talked to me and assured me, "Yes, you may be having some difficulty, but one thing I want you to realize is you're not the only one having difficulty." I had kind of felt like I was standing out. He said, "You'll be surprised that some of those you consider very articulate are not doing well." And he told me, "Hang in there. Stay with it. Just give it your best. I think you'll be pleasantly surprised." I did. I still got a C in that course, but it was a turning point for me.

> *It sounds like the burden of race and gender was a heavy load for you. Did you feel that your experience at an HBCU prepared you well?*

Dunston: I had the experience of being told, "Well, I don't know what these grades from Tuskegee mean." I mean, I took upper-level undergraduate courses as a graduate student because it wasn't clear what my grades meant in terms of Michigan graduate courses. I wasn't at the top my class, but I wasn't at the bottom of the class, either. Some classes I made A's, some B's, but it was a good experience. I didn't have any personally devastating academic experience that was not, in my mind now, possible for any graduate student. I remember my first seminar and being nervous. I suppose any graduate student coming from far away would have been nervous with their knees knocking. I got a positive response from my fellow students and the faculty. I was just glad that it was over.

Recently I was invited to serve on a National Academy of Sciences review committee and the chair of that committee was one of my instructors at Michigan. We reminisced about my days at Michigan. Two of the other committee members were at Michigan during the time I was there. It was almost like a homecoming.

> *Did you then go on to do a postdoc after finishing your doctorate at Michigan?*

Dunston: No, I didn't take that path exactly. I was really tired toward the end. I had literally gone from kindergarten to Ph.D. with no stops. Michigan was not a bad experience but it was tiring. For me, I was tired of seeing no blacks on a large scale. I had come from the South and I had been in schools where there were role models. There were people there whom I could identify with. I wanted something that I identified with to be a part of my world. Human genetics is a small program. I didn't have a lot of interactions, and I just wanted some more blacks in my environment. When I was at Michigan, we had one black seminar speaker and one other black student in my program.

Is that why you selected Howard University after graduate school?

Dunston: Yes, that is part of the reason. I'll tell you the story. I was at a scientific meeting and there were three black guys sitting around and joking and I was sitting at a small table that had a divider and in a very light-hearted way one of them said to me: "You keep looking over here. You know you don't want to be over there by yourself. Come on over here with us." They were at the meeting, I could tell. I went over and we introduced ourselves. It turns out that one of them was the chair of the microbiology department at Howard University. He had just taken the chair that fall. He wanted to build a microbiology department that had quality black Ph.D.s. He wanted to develop a research program. That was his goal. So he was "beating the bushes" for blacks to come to Howard to build a microbiology program. That guy was Willie Turner.

So you finished the program at the University of Michigan and went off to Howard University to help build a microbiology program.

Dunston: I finished my Ph.D. at twenty-eight, and I just wanted some positive black men in my life. I wanted some possibilities, and there had not been many at Michigan. These guys were so great. First of all, I was so impressed that they were just obviously bright and they were scientists. I also liked the way that they treated me at the meeting. They just took me under their wings at that meeting. They made me a part of their group. We went to dinner and we talked about what they were doing in science. In the meantime, Dr. Turner was great because he was trying to recruit me. He told me, "If you come and help me build this program, once we get you in position, I will assist you in getting you together with someone that you can do a postdoc with at NIH." So I took my position at Howard before I did my postdoc.

You took the position at Howard University but you did postdoctoral training before you began your tenure at Howard?

Dunston: Yes. The whole emphasis was moving to getting more blacks into Ph.D. programs and research careers, so here comes the MARC program, Minorities Access to Research Careers. They have postdoctoral grants, so I applied for one. Dr. Turner introduced me to the person at the Cancer Institute that indicated an interest in immunogenetics. I had done my Ph.D. from the genetics perspective, so I wanted to do the postdoc from the immunology perspective. I wanted some experience with the benchwork aspect of immunology and I was interested in cancer, so I met Dr. Herberman, who was head of the laboratory of immunodiagnostics at the Cancer Institute's immunology section. We worked out a project for me. I applied for the fellowship and I then took the time to do the postdoc. So I took leave from Howard University to do the postdoc.

At Howard, have you experienced any race and/or gender issues?

Dunston: Gender, yes. Howard is still a very chauvinistic place. I think there have been no racial issues because Howard is a predominantly black institution.

Can you elaborate on some of your experiences of gender discrimination at Howard or other places during your science career?

Dunston: I did not know it at the time, but my salary as an assistant professor initially was less than what the men received. This position at Howard was my first real career appointment. Prior to this, I had lived on a graduate student stipend supplemented by part-time work. I was so happy with the salary that I was offered, I never questioned whether it was equitable with others in my position. I progressed steadily in the position and was pleasantly surprised a few years later with a substantial increase in salary when the university took actions to adjust all salaries in line with the position. This resolved the gender differences in pay.

How many women are on your faculty?

Dunston: At the time I came, Dr. Turner was recruiting for new scientists. Our approach was to train our own and bring them back. We have thirteen full-time faculty members. In terms of recruiting, Dr. Turner was out really to get the best and brightest team of African Americans, people who were unique in both quality and capacity. In addition to me, he recruited three other women: Dr. Austin, a medical mycologist; Dr. White, an immunologist; and Dr. Gravely, a parasitologist. We have since added Dr. Day and Dr. Harris, whom we trained. The rest are males. He brought together a broad spectrum of disciplines both in terms of composition and training. When I came, we only had a master's program, so the first thing we worked on was getting a Ph.D. program in microbiology. With a Ph.D. program obviously we can have higher standards and we attract students that are investing more. Two of the females trained here are back here on the faculty, but the others are either in research or government-related positions. I think our students have gone on to careers that are reflective of our training. Howard certainly has its career differences in terms of women as you move up. I am now chair of this department because the chair was promoted to dean, and he asked me if I would serve as interim chair.

Tell me about some of the research that you've done here.

Dunston: As I said, my area of training is in human immunogenetics, looking at the genetics of the immune system. My interest is in a set of genes called the major histocompatibility complex genes and these genes came to recognition because they code for structures on the surface of the cells that are important in matching donor organ and recipients for transplantation. They are essentially the same as what we are testing for when we do tissue typing.

We've since learned that these same gene products that we call histocompatibility antigens are also intimately involved in regulating the immune response. The immune response plays a major role in host defense and maintaining biological integrity. I was interested in what's the basis for the differences between individuals in their susceptibility to various diseases. My research has focused on the identification and characterization of major histocompatibility genes and antigens in African Americans. Specific attention has been directed to the biomedical significance of genetic variation in these molecules.

And you've been working on this for twenty years?

Dunston: Well, different aspects of it. I came here and there was no one to really work with on the genetics side at Howard. I was first rooted in the teaching program and trying to get and build a Ph.D. program. Primary concerns were getting the kind of student who could do the work and at the same time trying to get financial support for the research. After I did the postdoc, I established collaboration with investigators within and outside of Howard. I got my first NIH-funded R01 grant, which was in cellular immunology because I had done the postdoc in tumor immunology.

When I got my first independent R01 research grant, I was the only one at Howard doing this type of research. I had to first set up a laboratory and recruit and train staff to do the work. I was given research laboratory space in our new hospital. Much time was required to inform the clinicians of the research project and to solicit their cooperation in obtaining tissue specimens for the project. Initial progress was compromised by the combined responsibilities of setting up a research laboratory, recruiting and training staff, developing collaborative research procedures with clinicians, and teaching and training graduate and medical students. Unfortunately this did not lead to a level of scientific productivity required for competitive renewal of the grant. After successfully setting up the lab and training the personnel, the grant was not renewed.

Is this the dilemma of being in an HBCU, or is it just starting out as a young scientist?

Dunston: It was a combination of factors, but it was another learning experience for me. Loss of the grant was devastating when I thought I had done great. I had set up all this ground work and the research protocols. I was giving talks about what I was doing, and now there's no money. What do you do when you lose your only grant support? The school certainly at that point didn't have an infrastructure to provide comparable support.

So I went back to the NIH lab for a year as a visiting scientist to stay active and to see which way to go with the work from there. The lab was pursuing some genetics aspects of tumor immunology, which was getting closer to my interests. I was fortunate because of another collaboration that I established, an immunogeneticist at Georgetown University was looking for someone to collaborate with so she could continue studying antigens in African American.

I love your determination to do research. Many people would have just given up. May I ask, was your new collaborator a white female scientist, and did the relationship prove to be fruitful?

Dunston: Right. Dr. Johnson is a white female scientist and the collaboration was scientifically and institutionally advantageous. At the same time Howard had a comprehensive Cancer Center that was supposed to foster collaboration with Georgetown. So from a university perspective, collaboration between investigators at Howard and Georgetown was encouraged. I was interested in HLA (human leukocyte antigen) and here was a new investigator that had done postdoctoral work and had done her training in HLA looking at African Americans, so we teamed up. With her experience and her contacts she told me that if I assisted her in getting resources to continue the work that she was doing she in turn would assist me in getting a laboratory set up at Howard to conduct research here. We applied for an NIH contract for reagents to better define transplantation antigens in blacks. We worked through the issues of getting appropriate samples and reagents to do the testing for the African American population. One of our ideas was to use mothers who had multiple births as a source of antibodies to HLA molecules common in blacks. We wanted to broaden our panel of HLA typing reagents to reflect differences in blacks and whites. Mostly, whites have been used as the source of reagents for tissue typing. Genetic differences between European and African populations require that resources for tissue typing be broader for African Americans than those commonly used for whites.

How did you select participants for this research?

Dunston: We set up a screening program. We approached the people in our black community and told them that we were trying to broaden the base of reagents used for tissue typing. HLA antibodies are routinely obtained from the serum of women that have multiple children. We ask them can we test their serum for antibodies to HLA.

Were the black patients and community open to that kind of testing?

Dunston: Yes. When you explain what you want to do and why, black patients and the community are very responsive. They are more open now because of all the things that have happened with genetic testing and screening. People want to have a lot more information and know it's not going to cost them. I mean the women are not going to be harmed. So, there was no cost physically or financially. Eventually, we didn't even have to draw blood because we could get serum from the placenta which was going to be discarded anyway. So we just had to get permission from them to test the serum for HLA antibodies. That's how we got started with the HLA work at Howard.

This lab was setup as a result of your collaboration with the scientist at Georgetown or you did you just expand what you had already started?

Dunston: Yes, in part. The contract really was a joint venture with the primary money going to Georgetown because the resources to define the specificities of the HLA in the serum were there. The Screening Program was set up at Howard but the laboratory analysis was initially conducted at Georgetown. We could then use the antibodies for tissue typing at Howard. The collaboration with Georgetown gave me a connection with the human immunogenetics scientific community. I had a couple of students in this program that did projects in collaboration with the lab there. We would train over there and do studies here. I joined the American Society for Histocompatibility and Immunogenetics, started going to regional and national meetings. I became more knowledgeable about current issues in the field. I then applied to the National Institutes of Health (NIH) for a capacity building grant to set-up a human immunogenetics laboratory at Howard. This was in 1985. It was probably in 1984 when the grant was submitted. Howard was awarded one of the first Research Centers in Minority Institutions (RCMI) grants. One of the components of our institutional award was money for an immunogenetics lab to facilitate HLA related research at Howard so the collaboration with Georgetown was instrumental in providing the groundwork to compete for RCMI support to set up the research laboratory at Howard.

Do you feel your professional organizations provide adequate opportunities. Do you think the issues that concern black women scientists are being addressed in these scientific societies?

Dunston: In the sciences, we're dealing as much with race as we are with gender. There are lots of women in the laboratory. As you move up to the higher-level positions, you start seeing the stratification between men and women. If you look at overall membership you have just as many women, if not more, in the audience. However, when you look at who's the director of labs, who's head of various science programs, that's when you start seeing gender differences. My organization, tissue typing group, has always had a really good representation of women. It emerged from a technology service aspect, so women were involved in forming the organization. There are no black women in leadership positions in the society. I am the only female director of an accredited laboratory. Our lab is the only accredited laboratory in a historically black college.

So you are in a unique position as probably one of the few women who head a science department in a research university. You are also director of a specialty lab.

Dunston: Yes. Our lab was accredited in 1990. I'm the only black director of one of these types of laboratories anywhere. There are other blacks that head tissue typing programs in majority schools. The charge has been for me to try to bring this program along and make it the best. That's been my personal choice and challenge. There are blacks that are higher in the field of transplantation immunology. They tend to move into the field from immunology or from the medical profession as opposed to genetics. I came to this discipline from human genetics. I've been going to meetings for almost fifteen years now, and the number of blacks at the top doesn't seem to change. You just don't see blacks in the pipeline. You see a few. It's vitally important that I note that.

Where do you see yourself in five or ten years?

Dunston: I want to still be doing research in human genetics. I am just fascinated with my discipline. When it comes to genetics, African people are just wonderful because we have the oldest history, and that's fascinating to me. This is a story in progress—we don't fully understand, but we are now recognizing the variability at the genome level itself. The gene was only a concept when I was in graduate school. Now we are mapping and isolating genes for study. As a child I looked at people and saw how we are different, and now I start looking at the nucleotides. We find that the diversity is staggering—how different everyone is. It's amazing that we share as much as we seem to share at the surface level. We know that this diversity plays a role in the susceptibility of a person to diseases and how the body regulates exposure and outcomes, etc. and that's what's so exciting. The challenge for the twenty-first century is to now read the "book of life" and learn who, what, and how we are as humans.

> *May I interject here? Your enthusiasm is absolutely beautiful, but how do you get all this work done with all your other responsibilities? How do you balance family and career?*

Dunston: Well, I don't have an immediate family here. I've mothered through my nieces and nephews and extended family. So in one way family has not affected my career. But you need balance and you have to work at that. I am a cancer survivor and I know that. I love my work and sometimes I have been discouraged with all the challenges, but I have been mostly pleased with how I live my life. What was the other part of the question?

> *How do you get things done with so many challenges?*

Dunston: Well, you have to think broadly and creatively. My challenges are with adequate funding for my current efforts to establish the National Human Genome Center at Howard University. As was the case with setting up the Human Immunogenetics Laboratory at Howard, we now need reference resources from African Americans and the African Diaspora populations.

> *What is your final advice to young black women wishing to become research scientists?*

Dunston: Determine what subjects or things attract your attention, what intrigues or captures your imagination. Learn all you can about the subject. Seek out people who share your interests. Identify a mentor, someone who exemplifies that which you would like to do. Develop the skills and talents you need to do the kinds of things you enjoy. Never question your ability to succeed. Be confident in your capacity to realize your dream and go for it!

Editor's note: Since this interview in December 1996, Georgia Dunston has established a laboratory for the human genome project for African Americans.

Selected Publications and Research Activities

Rotimi, C., G. M. Dunston, K. Berg, O. Akinsete, A. Amoah, S. Owusu, Acheampong, J., K. Boateng, J. Oli, G. Okafor, and B. Ostimehin. 2001. In search of susceptibility genes for type 2 diabetes in West Africa: The design and results of the first phase of the AADM study. Annals of Epidemiology, 11(1): 51–58.

Berka, N., G. N. Bland, D. P. Gause, P. F. Harris, L. H. Erabhaoui, A. S. Bonar, D. Okeapu, and G. M. Dunston. 2000. Early age of disease onset in African American type I diabetes patients is associated with DQBI*0201 allele. Human Immunology, 6: 816–819.

Mefford, H. C., L. Baumbach, R. C. K. Panguluri, C. Whitfiedld-Broome, C. Szabo, S. Smith, M. C. King, G. M. Dunston, D. Stoppa Lyonnet, and F. Arena. 1999. Evidence for a BRCA1 founder mutation in families of West African ancestry. Am. J. Human Genetics, 65: 575–578.

Hill, A. V. S., A. Sanchez-Mazas, G. Barbujani, G. M. Dunston, L. Excoffier, J. Hancock, J. Klein, U. A. Meyer, A. G. Motulsky, S. Presciuttini, W. L. Wishart, and A. Langaney. 1999. Human genetic variation and its impact on public health and medicine. In Evolution in Health and Disease (ed.) Stearns, S. C., Oxford University Press, Inc., New York.

Agurs-Collins, T., K. S. Kim, G. M. Dunston, and L. L. Adams-Campbell. 1998. Plasma lipid alterations in African American women with breast cancer. J. Cancer Research & Clinical Oncology, 124: 186–190.

Dunston, G. M., et al. 1997. Collaborative research on the study of the genetics of asthma (CSGA), a genomic-wide search for asthma susceptibility loci in ethnically diverse populations. Nature Genetics, 15: 389–392.

Ofosu, M. H., G. M. Dunston, L. Henry, D. Ware, W. Cheatham, A. Brembridge, C. Brown, and L. Alarif. 1996. HLA-DQ-3 is associated with Graves disease in African Americans. Immuno. Invest., 25: 103–110.

Berka, N., G. M. Dunston, B. I. Freedman, D. Gause, G. McCintoch, and A. Alim. 1995. Absence of HLA-DR3, DQ polymorphism in African American patients with end stage renal disease. Human Immunol., 44: 79.

Callender, C., L. Hall, C. Yeager, J. Barber, G. M. Dunston, and V. Pinn-Wiggins. 1991. Organ donation in blacks: The next frontier, New England Journal of Medicine, 6: 442–444.

Dunston, G. M., W. L. Henry, J. Christian, M. Ofosu, and C. O. Callender. 1989. HLA-DR3-DQ heterogeneity in American blacks is associated with susceptibility and resistance to insulin dependent diabetes mellitus. Tran. Proc., 21: 653–655.

EVELYN BOYD GRANVILLE

Destined to Greater Heights

BORN IN WASHINGTON, D.C. IN 1924, Evelyn Boyd Granville grew up during the 1930s. Although the Great Depression of the thirties devastated families across America, she could not remember being poor or without the necessities of life. After her father left the family, her mother and aunt assumed the sole responsibility for raising Evelyn and her sister. Her mother found employment at the United States Bureau of Engraving and Printing and worked there until she retired. Though Washington was quite segregated at the time, Evelyn and her sister found life to be rather pleasant in the capital city. Unlike many places in the Deep South, the libraries and museums were open to all and provided many hours of learning and enjoyment for Evelyn. She loved school, espe-

Evelyn Boyd Granville, one of the first black women to receive a doctorate in mathematics in the United States, is a professor emerita at California State University.

cially mathematics. She recalls feeling that the schools for "coloreds" were in no way inferior to other schools: "I truly felt that no teacher during my entire K–12 education didn't demand excellence and they were dedicated to educating us at the most superior level." She graduated from the famous Dunbar High School, which had a history of graduating some of the most outstanding black leaders during that time. The school encouraged students to attend the Ivy League colleges of the Northeast. Evelyn was no exception.

In 1941, Evelyn entered Smith College as a freshman with help from a scholarship from Phi Delta Kappa, a national sorority of black teachers, and from her aunt, who paid part of her expenses. The largest private college for women in the United States, Smith can boast of other outstanding graduates, such as Betty Friedan, Jane C. Wright, and many others. In fact, Friedan, author of *The Feminine Mystique,* was a senior and the editor of the college newspaper when Evelyn entered Smith. Evelyn concentrated on her studies in mathematics, astronomy, and physics. She even toyed with the idea of becoming an astronomy major but decided to focus on her first love, mathematics. She was elected to Phi Beta Kappa and to Sigma Xi, both outstanding honorary societies. In 1945, she graduated summa cum laude from Smith College with honors in mathematics.

After completing her undergraduate degree, Evelyn was accepted at two highly rated graduate programs, the University of Michigan and Yale University. She decided to go to Yale University because of the fellowship they offered to supplement her funding from Smith College and because of the opportunity to study with Einar Hille, a distinguished mathematician in the field of functional analysis. Interestingly, her decision not to attend the University of Michigan resulted in her not meeting her sister in mathematical science, Marjorie Lee Browne (now deceased), who received her Ph.D. in 1949 in mathematics. Both women have been given credit as the first black women to receive doctorates in the United States. However, it has been confirmed that Euphemia Lofton Haynes was the first black woman to receive a Ph.D. in mathematics, from Catholic University in 1943. All three of these black women forged the path for other black women to pursue doctorates in mathematics. The first white woman to receive a Ph.D. in mathematics was in 1886 and the first black man to receive a Ph.D. was in 1925. It was many more years before a black woman could realize the dreams of her white female and black male counterparts.

Evelyn Boyd Granville has had a distinguished career in mathematics. She began her career as a research assistant for the New York Institute for Mathematics. From there, she accepted an appointment as an associate professor at Fisk University, in Nashville, Tennessee, but this only lasted for two years. From this experience, she knew that she liked teaching and that she preferred to live in the south. Washington seemed to be a suitable place, and she returned there in 1952 to work for the National Bureau of Standards as a mathematician. Several positions were to follow this one. When she got married in 1960, she moved to California, where she worked with North American Aviation on a project for NASA's Apollo Space Program.

By 1967, Evelyn had a wealth of experience with various projects and companies; unfortunately, she no longer had a marriage. She then had to make a crucial decision: whether to stay in private industry or seek an academic position. She chose the latter and never regretted her decision. At California State University at Los Angeles, she progressed through the ranks. Along with this new change came many exciting opportunities in teaching and mentoring the next generation in mathematics. In addition, she co-wrote, with Jason Frand, a mathematics textbook that was used in over fifty colleges in the United States.

In 1970, she married her second husband, Edward Granville, a successful real estate broker in Los Angeles. The Granvilles decided to retire in 1984 and moved near Tyler, Texas. Retirement, however, was short-lived for Evelyn. She decided to teach at Texas College, a predominantly black college, before moving on to the University of Texas at Tyler.

Although Evelyn doesn't teach at the college level anymore, she still is quite active in the community at large. She served as a consultant for a Dow Chemical science program for elementary and middle school teachers. Granville is a woman of her generation, yet she was an exception to the rule during her early years. She never thought about being poor and underprivileged but only about the destination her life could take. As she wrote in 1989,

> When I was growing up, I never heard the theory that females aren't equipped mentally to succeed in mathematics, and my generation did not hear terms such as permanent underclass or disadvantaged. Our parents and teachers preached over and over again that education is the vehicle to a productive life, and through diligent study, we could succeed at whatever we attempted to do.[1]

Clearly, race and gender were significant issues for black women, especially before the war efforts of the 1940s created educational and career op-

---

1. Evelyn Boyd Granville, "My Life as a Mathematician," *Sage, A Scholarly Journal on Black Women* 6 (1989): 55–58.

portunities. Yet Evelyn never worried about those issues because her family and community had supported and encouraged her to reach for her dreams. By the time she attended graduate school and began her career, she was a woman with high self-esteem who had the skills to succeed. It has been over fifty years since Granville made history as one of the first black women mathematics doctorates in the U.S. Her kind of experience in science should be the rule, rather than the exception.

How did you first become interested in science?

Granville: I loved science. It came naturally. I also loved mathematics. When I went to college, I studied astronomy and physics. Both of these subjects are mathematical in nature.

As a child, did your parents encourage you in science or notice something unique about your ability to do math?

Granville: No, nothing special. I was a good student and I got good grades.

Was there someone special who encouraged you early on in science? What role did your family play in your selecting science as a career?

Granville: No, there was no one special, but my family supported me in going to college. I grew up in Washington, D.C. We (my sister and I) just knew we were going to college. Well, the support that students received was enormous. You were expected to go to college. You got good training and support. You were strongly encouraged by your teachers. Even though it was a segregated system, you got excellent support. You got the cream of the crop in terms of teachers.

I noticed that you went to Smith College. Was there any particular reason that you went there?

Granville: I attended Dunbar High School in Washington, D.C., and we were encouraged to attend Ivy League schools. The young women were encouraged to apply to schools like Wellesley, Radcliffe, Mount Holyoke, Smith, and Vassar. And the young men would be encouraged to apply to Harvard, Yale, Princeton, Amherst, and Dartmouth. So it had been a tradition that many of us applied to the traditional northeastern Ivy League universities and colleges. My family was low-income financially but I was still encouraged to consider and apply to these colleges. My homeroom teacher, who happened to also be my math teacher, encouraged me to consider Smith College because her sister and her niece were graduates of that college. They all encouraged me to go there. I also applied to Mount Holyoke and I was admitted to both of those colleges. It was my homeroom teacher who pushed me to go to Smith College.

So Smith College had a tradition of accepting young black women.

Granville: Oh, yes. Another young woman who had gone there in the 1930s was Laura Cole Phillips from Washington, D.C. The year I graduated from high school several of us went to these northeastern colleges like Wellesley, Vassar, Radcliffe, and Mount Holyoke.

What were your experiences like at Smith?

Granville: I had wonderful experiences. I never had a problem as far as the color of my skin was concerned. I had excellent teachers and I never had any problems with the teachers or my fellow students. I lived in one of the dormitories on campus. Most of the young women were well-to-do, but others were, like me, at the bottom of the income ladder. I never felt poor or disadvantaged.

Were you the only young black woman at that time in science or math at Smith College?

Granville: Yes. At that time there were only about four young black women on the entire campus.

Was there any bonding or keeping in touch with them?

Granville: No. I don't remember having any contact. We had nothing to do with each other.

Why? Please explain that.

Granville: When I was a freshman, Jane Wright from New York was a senior at Smith.

Do you mean the famed physician, Dr. Jane C. Wright, from New York?

Granville: Yes that's her. Jane Wright, M.D., was there when I was a student. There was also Jane White, who was the daughter of the head of the NAACP, Walter White. We had no contact with each other. We didn't make a conscious effort to avoid each other. We just had no reason to be in touch with each other. Our paths didn't cross on campus. We were in different dorms and I was a freshman. I think Jane White was a junior when I first came there. I really didn't need to seek them out for anything.

That is quite interesting. I thought with such a small number of young black women that you would seek each other out.

Granville: As far as seeking out another black person on campus for support, it wasn't necessary because the college environment provided the support that I needed. One year, when I was a junior or senior, another young black woman transferred to Smith College and we lived in the same dorm and we were friends—because our paths crossed.

It wasn't because you were black that you did not make friends with other black students, it was because your paths didn't cross. Often a black person on a white campus is criticized by other blacks for not seeking out other blacks. At least, that was one of my experiences. You have provided an interesting exception to that experience. Of course, I think there is a generation or time factor that may account for this difference in experience.

Granville: Right. There weren't enough of us to get caught up in who was sitting with whom and assimilating solely based on race and gender. Now, if there had been more of us, then our paths might have crossed and there would have been more interaction.

It was a different time. I think your comments speak highly of Smith College. If Smith College was providing what you needed as a student and you didn't need to seek out other blacks, then that sensitivity to nurture was admirable of Smith.

Granville: Yes, Smith provided what I needed. I had my friends in the dorm. I had my friends and classmates who studied and dined with me.

You graduated from Smith College in 1945. You received a fellowship from Yale. Were there any other universities that you considered for graduate training?

Granville: Yes, I also got accepted to the University of Michigan. That's where Marjorie Browne attended. However, Yale offered me a scholarship.

Marjorie Browne shares the honor with you of being one of the first black women to receive doctorates in mathematics in the United States; did you ever meet or even talk on the phone?

Granville: No, we never met and I never knew anything about her. I am not sure she knew anything about me, either.

I am surprised to hear that. Did you eventually learn of each other?

Granville: I only knew of her when someone said that we were the first two to receive a Ph.D. in mathematics.

When did you find out you were one of the first black women to receive a Ph.D. in mathematics in the United States?

Granville: I didn't know it. I didn't have a clue until my sister said to me one day, "Did you know that you were one of the first black women in America to receive a Ph.D. in mathematics?" I said, "Oh, really." I didn't attach much importance to that because it had not been a goal of mine. It wasn't something that I was trying to achieve. To me, it was just incidental.

What happened once you left Yale? How did you choose the academic route?

Granville: Well, I had planned to teach in a public school system. But I didn't have the education courses, so that was out. The first year after I left Yale, I got an appointment at New York University as a research assistant. During that year I applied for a job at a college in Brooklyn, New York. I thought the interview went very well. Later on, according to Pat Kenschaft, a mathematician who also does research on women and minorities in mathematics, the interviewer at that job actually thought it was a joke that a black woman applied in the first place. I had not thought about it much until she relayed these facts to me. I wasn't aware of any animosity from that interview experience. I just thought that you don't get every job you apply for anyway. Frankly, I didn't like New York, and it was quite expensive, even in those days, to live there. So I wasn't disappointed. I met the president of Fisk University, and he invited me to come there as a professor. I thought it would be a good experience and Fisk University had a great reputation.

That leads me right into the next question on race and gender. Do you feel race and gender were used to define you in your career or that they were factors in how you were viewed as a scientist?

Granville: I viewed the reaction to my race and gender with amazement. You could see it on their faces. "What are you doing here?" was what they seem to be saying. The fact that I am a woman, African American, and a mathematician was something that I allowed to be the other person's problem. I guess I should clarify and say that it was not a problem as much as it was so rare in those days to see that combination. I don't think race or gender kept me from getting a job. I am not aware of that except where I was explicitly told, as in the case of the teaching job that I applied for early in my career in Brooklyn, New York.

In fact, I think my race and gender served as a highlight or advantage for me. You stand out if there are a lot of people applying for the same job. Whether or not your diversity also hinders you or someone uses it to hold you back or favor you, I am not sure. How a potential employer or person in a power position might use your race and gender to deter you is difficult to sort out. I don't think my race or gender influenced my getting or changing jobs. I never felt hampered by race or gender. I am not aware of whether or not something relative to race or gender was happening behind the scenes. I think it was a matter of timing, also. I came before the 1960s. In the 1960s, when we had the start of the black power movement, the push for equality, and the riots, then the attitudes changed. If I had been born twenty years later, I might be telling a different story.

> So you are saying that because you were a pioneer race and gender did not influence your career negatively. Your kind of developed talent and skill in our race was so rare that people were just amazed.

Granville: Yes, I think it helped me during that time. Time and place are important. Remember, this was in the 1940s. It worked in my favor. I was considered an exception and someone they had not seen before. Women in mathematics still have that stigma that they are not supposed to do math. It's still a myth. There were other women in math and in the graduate program when I started Yale in 1945. I think there were about five or six of us. Only two of us finished the Ph.D. program. The others got the master's degree. At Yale during that time in 1945, World War II had ended and the men were coming back to finish undergraduate degrees, and not graduate degrees.

> That's a good point about the influence of the war for blacks and women. In some ways the war effort really opened doors for African Americans and women. Would you please comment on that important part of history?

Granville: Yes. World War II opened up jobs for minorities that had never been open before. The government and private industry needed workers, and Washington, D.C., was a key example of that situation. I personally knew so many people who got government jobs. When I was growing up there, there were two government agencies that hired black women (in those days we were called "colored ladies"). One was the Government Printing Office and the other one was the Bureau of Engraving and Printing (where money and stamps were made). My mother and aunt worked at the Bureau of Engraving and Printing. These were good jobs but not very high-income jobs. You earned a decent wage and you didn't have to worry about being laid off. These were the two agencies where "colored women" were hired in large numbers. My sister started college in 1939 but quit to work with the Census Bureau. She stayed there for over thirty years.

> *I've read that once the men returned from the war that the women had to give those jobs up and go back to being homemakers. Can you tell me about that experience for black women?*

Granville: Well, that applied more to white women. Black women had to keep the jobs that they had. Remember that these jobs were not the top jobs. Black women were the heads of households and they had to work. Black women could get jobs even when black men couldn't because they were hired as maids, caregivers, child caretakers, etc. White families needed them in that capacity. They may not have been the best jobs, but they were jobs. The black man has been emasculated for centuries because he was considered a threat, but black women are not viewed as threatening and therefore they could get work. The defense industry was another area that the young black men worked in during the war. World War II made a big difference for people of color.

> *Did the civil rights movement or the women's movement play a role in your career at all?*

Granville: No. I was already there. It might have helped women coming along after me.

> *How do you feel other scientists can be more supportive of black women scientists?*

Granville: I guess by being a positive role model and mentoring. I think blacks in general have to help themselves. We have to deal with the problems in the black community ourselves. When I was growing up, our professionals lived in our community. We lived together and went to church together. Now you have the BUPPY syndrome (the Black Upwardly Mobile Middle Class); they live in the suburbs and not in the inner city or the black community. There is a disconnect in today's generation of blacks. I guess I am trying to say that support systems for blacks in science (male or female) must start early and within our community. As I said before, I was a disadvantaged kid, but I never felt like one because of the high expectations of my parents, teachers, and the black community. It was understood that I would be going to the best colleges in the country and that I would do well. The black urban kid doesn't even see a black teacher sometimes because black professionals are no longer in those communities. We have to find some way to interact with our children at a young age to let them know what's possible and what's expected.

> *So you don't feel that whites or other groups have a stake in supporting blacks and/or women in science?*

Granville: Well, why should they bother? What I mean is that the support must start with us. Please don't misunderstand me, I don't think that white people in this country don't have a responsibility to be supportive of diversity. White people should care about what happens to that urban kid for their own good. However, we blacks have to be the leaders in demonstrating that that inner city kid is worth saving and should be valued in this society. Unfortunately, stereotypical images (like drug dealers, gang violence) don't help whites to understand the needs of the urban community. While it is true that some of these images are distorted and are unfairly targeted at our youth, they cause white Americans to feel less sympathy.

> *What you are saying is true, but what about Columbine? The problems of the inner city youth are no longer just in urban areas; they have spread to the suburbs.*

Granville: Yes, I agree. The recent violence displayed by youth of America should be a disturbing phenomenon for white America and for all of us. After all, many of them left the inner cities for suburban life. We are losing our inner city youth and serious problems are brewing in our suburban youth. There is a lack of connection in both groups to our young people.

Back to your original question: a support system has to be built early for our youth and especially for our young black girls. Science and math are generally rigorous curricula for any youth, but long-term success depends on early preparation and interest. I think the greatest support has to come early on because this will ultimately have profound impacts on our scientific and technical workforce in the future.

> *Let's talk a bit about family. How have marriage and having a family played a role in your life as a scientist?*

Granville: No role at all. I married a minister the first time and that was a disaster. I was married for about seven years the first time and we also had his children from a previous marriage with us. That simply added to a disastrous marriage. It was not a good marriage and I got out of it as fast as I could. Although it was bad marriage, I felt it did not affect or influence my science career. The second marriage was very good because my husband was supportive of everything that I was doing. I think that is because he was a well-established and secure person in his own right. He was a real estate broker and had been in business for years when we met. He was not intimidated at all that I was a mathematician. Our relationship is a very good one because we enjoy many of the same things, like traveling together. I married him in 1970 and I was well-established in my career. So he had no particular influence on my science career.

> *Are there any particular strategies that you would recommend in balancing a science career and family?*

Granville: I don't think that there are any differences in balancing a science career than any other career. Women in the workplace who are mothers have got to plan. Parents have to plan in any career on how they are going to be supportive of their children and they have to really work on building a strong family. I think that's true irrespective of what field you are in. More than ever, parents are going to have to start being parents in the truest sense so those disastrous situations like Columbine don't happen. We have so many "latchkey" kids. I was one myself because my parents were divorced. My mother worked, but we knew to come home and do our homework and chores.

> *What made the difference with your generation?*

Granville: Expectations and what our parents would tolerate. I would never sass my mother the way I see children do these days. Parents would support you, but they expected and demanded respect.

> *So you think parents simply tolerate too much these days?*

Granville: Yes. I think parents have to start parenting whether they are working or stay-at-home parents. A lot of stay-at-home moms don't parent, either. Parents give their children too many things. There are too many diversions.

> *You seemed to imply earlier that women don't have to make choices between having a science career or another career.*

Granville: I think women who want it all have to realize that they might have to make some sacrifices as far as time is concerned. Perhaps, while the children are growing up, they might want to find a job that does not require as much time. They might want to find a job that would permit them to work in their chosen profession but at the same time not take the work home with them in the evenings or weekends. Of course, for the academic positions that's a problem. The woman professor knows she's got to write, publish, produce, and get money for research projects. She might consider not working in academia until later and opt for an industry job that might be more conducive for a family lifestyle. Then the pressure of tenure and promotion would not complicate her life as much. Yet there have been some women who have had children and been quite successful in the academic world. In those cases, they needed to pick a husband who was going to help them. The woman has to really use her head to pick a mate to support her. She has to make sure he is on the same wavelength with her to properly raise their children and have a strong family.

Well stated. Tell me a bit about your research days.

Granville: I worked on a variety of research topics both in industry and academia over the years. At IBM, I worked on developing programs for the IBM 650 computer. We were responsible for writing computer programs to track the paths of vehicles in space. The work was quite interesting. I also worked for NASA on the Apollo Project. I did research on trajectory analysis and orbital computations for NASA's program.

During my academic career, my focus shifted a bit to developing more effective teaching methods for students and teachers. I taught in several summer programs during this time for school teachers. It was about improving their teaching skills in the classroom. Jason Frand and I co-wrote a college textbook used for training elementary school teachers.[1]

What can black women do to change our invisibility in science?

Granville: As scientists, you have to attend conferences in your discipline. You can also do things locally, like school or science fair programs. Mainly, you need to be a part of your professional scientific societies.

What's next on your agenda?

Granville: In 1998, I was invited by a public relations company in Houston for Dow Chemical to use my picture on an ad campaign for pioneers in science. That led to my being asked to visit middle schools in Texas to speak about math preparatory skills. That began in February 1998 (Black History Month). I traveled to Houston, Dallas, and Beaumont, Texas, and Shreve-

---

1. Jason Frand and Evelyn Boyd Granville, *Theory and Applications of Mathematics for Teachers* (Belmont: Wadsworth Publishing, 1975).

port, Louisiana, to talk to elementary and middle school students, parents, and teachers about the importance of studying mathematics.

> *Where do you see yourself in ten years? You are so active and you really haven't truly retired.*

Granville: I will be doing the same thing, hopefully. I want to inspire young people to study math.

> *What do you think your greatest contribution to science will be or is?*

Granville: I don't think that I've made any great contribution to science at all. I think I see myself as a person who had a talent for mathematics who got the education to do the job. The education enabled me to work in interesting jobs. There were no computers when I was growing up, but because of my training and math background, I was able to join IBM in 1956 and learn computer technology. I used that training to work and generate programs for the NASA space programs. I would like young people to realize that education opens doors for you. The study of mathematics is really essential. Education is the most significant indicator of success in business and the professions.

There were two young women that I taught at Fisk University that I believe I had an influence on: Vivenne Malone Mayes and Etta Zuber Falconer. If I made any contribution to science, then it would be that I inspired them to become mathematicians. They both have doctorates in mathematics. I guess that's the most any of us can hope for: to live our lives in such a way that it inspires others to achieve. There's no greater contribution that one can make.

> *That is so true. What final advice would you like to offer young women?*

Granville: Go for it. Identify your talents and develop them. Get the education that helps you use your talents wisely. Don't forget to get a broad education. I tell students that it's great to be a scientist, but they've got to remember that scientists have other talents than just knowing science. Be prepared and go for it.

Interview Date: October 1999.

Editor's Note: Since the time of the interview, we learned that Dr. Euphemia

Lofton Haynes received her Ph.D. in mathematics from Catholic University in 1943. Thus, Dr. Granville and Dr. Marjorie Browne (deceased) share the honor of being the second black women receiving the Ph.D. in mathematics in 1949. (Personal discussion with Dr. Granville and Black Women in Mathematics through the *Mathematical Newsletter* website).

SHIRLEY ANN JACKSON
The Sky Is the Limit

BORN IN WASHINGTON, D.C., IN 1946, Shirley Ann Jackson credits her parents as being supportive and strong role models in her life. Her strong interest in science and mathematics was nurtured by her family and teachers. She was selected to be in the honors program in the seventh grade, and by her senior year in high school she was thoroughly prepared for college-level courses. Jackson was encouraged to apply to the Massachusetts Institute of Technology (MIT) by the assistant principal at Roosevelt High School. After graduating as valedictorian of her class in 1964, Jackson enrolled as a freshman at MIT in the physics department. This set her on the path to a great career.

In 1964, the first civil rights voting legislation had passed which gave Afri-

Shirley Ann Jackson was the first black woman to earn a doctorate in theoretical physics and the first black and woman to head the United States Nuclear Regulatory Commission.

can Americans the right to fully participate in the American political arena. When Jackson arrived at the highly regarded MIT, she was one of about ten African Americans in a student body of about eight thousand. About forty-three women were in her freshman class of nine hundred. Jackson felt isolated as an undergraduate student, but being the confident, self-assured, and well-prepared person that she was, she was undaunted. In 1996, she humorously related, "I can still recall the day in the mid-1960s, when I was deciding what to study in college, being told by a professor that 'colored girls should learn a trade.' Needless to say, this caused me a great deal of angst, but I responded by choosing a trade—Physics!!" Jackson says that she was able to develop friendships, "and people began to judge me on my merits."

Jackson received her B.S. in physics in 1968 and continued graduate studies in the field of theoretical elementary particle physics. She received her Ph.D. in 1973 from MIT under the tutelage of James Young, the first tenured black professor in the physics department.

Jackson spent three years at various labs furthering her expertise in research. Her first postdoctoral year in theoretical physics was spent at the Fermi National Accelerator Laboratory (1973–1974) and the next year was spent at the European Organization for Nuclear Research in Geneva, Switzerland, followed by an additional year at Fermi.

Jackson joined the theoretical physics staff at the A T & T Bell Laboratories in 1976 in New Jersey. She worked there for 15 years and conducted research in theoretical physics, solid state and quantum physics, and optical physics. Her outstanding contributions in physics have earned her significant honors: she is a fellow of the American Academy of Arts and Sciences and of the American Physical Society and has been granted several honorary doctoral degrees from various universities. Jackson has received New Jersey's Governor's Award in Science (the highest honor to a state citizen), was inducted into the National Women's Hall of Fame, and in 2001 was named Black Engineer of the Year.

Jackson, like some other women scientists, married a fellow physicist during her career at Bell Laboratories. They have one son. Jackson describes how important it is to have a supportive spouse and family.

From 1991 to 1995, she was a professor of physics at Rutgers University before being appointed as a commissioner to the Nuclear Regulatory Commission in May 1995. In July 1995, she became chairman and served until 1999. She became the first woman and first African American to serve in this role.

Jackson has had many firsts in her life, and these achievements continue. In July 1999, she became the first black and first woman to serve as president of Rensselaer Polytechnic Institute, one of the oldest technology institutions in America. She has also being selected as president of the American Association for the Advancement of Science. Shirley Jackson is not a person who is afraid of the possibilities and challenges of reaching for the stars. In her case, the sky is truly the limit.

From my background reading, I learned that your father had a big influence on encouraging you and on your selection of science as a career. Was there a particular influence that your mother had, or was it strictly through your father?

Jackson: No, both of my parents had a big influence on me. First of all, they were very oriented to education and higher education. My mother was very strong in language arts and taught us to read as children. It was because of her that I have the ability to write reasonably well. And I've always loved to read.

Your biography mentions that you have two other sisters. Are they also in science?

Jackson: No. One of my sisters is a lawyer and the other is a historian and university dean.

Was there a particular role that you felt teachers played in your selecting science?

Jackson: I think that when you put the question that way, it's hard to answer. I would say that I probably had a natural interest in science. The role that my teachers played was one of encouraging and nurturing that interest in science and math. I had a particular interest in mathematics, which is probably why I'm a theoretical physicist.

Where did you grow up?

Jackson: I grew up in D.C.

Did geographical location have any influence on your selection or interest in science?

Jackson: I was a child of the Sputnik program. After Sputnik, Washington put into place a somewhat controversial tracking system. I was in an accelerated program where I essentially finished most of my high school requirements by the end of junior year and took somewhat more advanced subjects when I was a senior in high school. I'm sure that was a kind of reinforcing situation and I felt that when I did go to college, I was pretty well prepared from an academic point of view.

> *In your biographical sketch, it was stated there were twenty African Americans and forty-two women in your freshman physics class.*

Jackson: There were about ten African Americans at Massachusetts Institute of Technology. There were about forty-three women in my class, including one other African American female. There were two of us.

> *Do you feel race or gender played a particular role in how you were viewed as a physicist?*

Jackson: Yes, probably still at this stage, but less so than when I was starting out. When I was an undergraduate is probably when it had its greatest influence. Whites were not accustomed to seeing someone who looked like me in those classes. In some of the classes, there was isolation and so on, but I found that as time went along, people knew I was serious and I was good at what I was doing, so gradually things started to change.

I think today because of who I am and where I am, it has less of an influence in the negative sense. But I think there's a sense in which one can use gender to her advantage; when I am at international and national scientific meetings, being an African American female, people tend to remember me. So when I presented my work, people remembered my work and associated that work with me. When one tries to do a good job and present some interesting results, it can actually be somewhat helpful. But I'm not implying that things were all peaches and cream all the time. I think not. There were hard times, but it related back to my earlier undergraduate years. I've been out of graduate school over twenty years.

> *Would you say race or gender had more impact? Can you separate them out?*

Jackson: Well, I think it depends on the circumstances. It's not always easy to separate. Probably in the early days because of when I went to college, race may have had a bigger impact. I think that gender was always there. I think later on the gender aspect became more apparent. I decided the only way that I was going to succeed was to focus on physics and not on my race or gender. I let that be somebody else's problem.

> *This is a beautiful response for our young people to hear, especially at the graduate level. In other words, when you become well-prepared*

in your discipline and focus on doing your best, even when you are confronted with race and gender issues, that doesn't have to deter you from your goals. What role did the civil rights or women's movement play in your success in physics?

Jackson: The civil rights movement and the women's movement helped to ensure more doors were open and provided opportunity. They gave me more confidence.

You mentioned that you joined a black sorority and that was helpful. Which sorority?

Jackson: Delta Sigma Theta Sorority. That was a social outlet. In addition, the Deltas did volunteer work in the community. We worked in Roxbury, Massachussets, which was a predominantly African American area. It was important because it put me in touch with the African American community. I've always believed that if one has talent and opportunity, one has to try to do something for more than oneself, and that was my objective.

You spent a year in Geneva, Switzerland. How was that experience?

Jackson: Oh, I enjoyed it. I was there doing research in the theoretical division for a year. I had a fellowship. It was fun living in Switzerland and doing a different kind of research.

Do you participate in professional organizations, and do you think they offer a service or meet the needs of African Americans?

Jackson: I belong to all the major professional organizations related to my field. I am a fellow of the American Physical Society, which is the primary professional society for physicists. I am also a fellow of the American Academy of Sciences. I am or have been a member of a number of other professional organizations, among them the Optical Society of America, the Materials Research Society, and the American Association for the Advancement of Science. The National Society of Black Physicists allowed African American physicists and related scientists an opportunity to get together and talk about research. I think it was particularly useful and encouraging to students.

I'd like to shift the topic a little bit. What impact, if any, has your family had on your career, and was it positive or negative?

Jackson: My family always has been absolutely positive.

Your husband is a physicist?

Jackson: Yes, he is a physicist.

So he's helpful in understanding the issues. What do you see as your greatest challenge as head of the Nuclear Regulatory Commission?

Jackson: To ensure the safety of various nuclear facilities and activities that we're responsible for regulating. To remain open to the public and help the public to understand how we do our job, and to be fair to those we regulate. The one enhances the other, so it's a set of intertwined issues.

What do you enjoy most about your job?

Jackson: The fact that it is a multifaceted job.

In what ways?

Jackson: Well, I am a member of the Commission and chairman of the Commission, where I am essentially working on policy formation and rule making and adjudicating, so there are those kind of collegial functions. But I am also the principal executive officer and I am the official spokesperson for the Commission. I have testified before Congress. As the principal executive officer, I am also the one who oversees what the staff does. I enjoy that. The interaction with the Congress, the international aspects, and interacting with the public are all things that I enjoy doing with this position. So it's a job that has many aspects.

It sounds like it is always exciting.

Jackson: Oh, yes. And then it has its technical undercurrent, which I like, but I don't get specifically into calculations these days, because this job is mostly policy and management.

What is your vision for the organization and your goals for your tenure? This is an appointment by the president. How long will your appointment last?

Jackson: Right, it goes until 1999. I've had an operational vision. It has three elements: reaffirming our fundamental health and safety mission, enhancing our regulatory effectiveness, and positioning the NRC for change. The environment in which we operate is changing because of initiatives in Congress. Change is happening to those we regulate, as well as changes internationally. We may be called upon to take on additional responsibilities. For example, external regulation of the Department of Energy may be one area. So we are having to look at how we do our work, and why we do it, and think more about it. In addition, we're trying to do more performance-oriented regulation and to take into account a risk-informed perspective.

I think you've touched on this, but I would like to hear more. I believe role models and mentors are important, especially for black women scientists. How do you see role models?

Jackson: I think role models and mentors are good if you can get them. I think they're not always there. I tend to think of a role model as more than what can actually be embodied in one person, but it can also be an avenue or a path to get from one place to another. I think one can have unwitting mentors—that is, not someone who necessarily takes a direct and personal interest in you. But given a particular situation, this person may decide to help you because you are in the right place at the right time.

I think that for many African American women things happen that way. I can't say I had a real direct role model. My parents in some sense were. They were not scientists. But in the sense of how they lived their lives, they were role models. Their values, their focus on education, their support for their children were all positive aspects of what role models should be.

So you wouldn't say that you had a particular mentor in your career as such?

Jackson: Various people influenced me and helped me along the way. I don't know that I had somebody who took me under his or her wing, told me how to shape my career, but I had a number of people who helped me.

How do you think black women can become more visible in science or engineering? Do you think it's important that we become more visible?

Jackson: I think that it is important for people to build successful careers. However, that requires a good grounding and background in one's discipline to be able to pursue a particular career as well as take advantage of the opportunities that may present themselves in a particular discipline. It requires that the person decide what she wants to focus on. Success can come in various ways—recognition of one's research, teaching, service. I think that each person has to decide what success means to her. In general, I think visibility is important. The issue, then, is to look for an opportunity.

What do you think is the most important advice you can give a young African American female in science? I talk to young African American women, and they often say they do not see someone like me or you in nontraditional fields.

Jackson: Well, I think for a trailblazer there is always a certain loneliness. I think one is going to go through certain things whatever career one chooses.

The difference is that there may be others around the person who maybe look like her—but she still has to work hard. There are going to be certain struggles. I enjoy what I do. I've been through many things but I don't talk about all that happened to me. I've been through a lot. In the end I am where I am.

So it makes it all worthwhile. You did research when you were at Bell Labs? Did you also hold adjunct positions?

Jackson: No, I didn't become a faculty member until I went to Rutgers. I came as a tenured full professor.

Why did you choose Bell Labs, or did they choose you?

Jackson: I would say some of each. My degree was in theoretical physics, but I was always interested in other aspects of physics. Bell Labs at the time was one of the best places in the world for me. In those days, Bell Labs would hire a person they thought was smart and could contribute to science and the organization. So when Bell Labs offered me the opportunity to shift my focus from particle physics to condensed matter physics, I took them up on it.

What do you think your greatest contribution to science will be?

Jackson: My work on the electronic and optical properties of two-dimensional systems, especially polaronic and related aspects of electrons on the surface of liquid helium films. I also believe my work in science, technology, and public policy—culminating in my appointment by President Clinton as the Chairman of the U.S. Nuclear Regulatory Commission. I also believe my work in changing the regulatory approach at the NRC to an informal, performance-based regulation will be viewed as important.

Is there anything else you'd like to add?

Jackson: Reach for the stars. It always gives you an upward trajectory. If you do not reach, you do not get anywhere.

Interview Date: December 1996.

LYNDA M. JORDAN

An Unlikely Scientist

ON FIRST EXAMINATION, LYNDA JORDAN might appear to be an unlikely person to become a research scientist. Lynda was born in 1956 in a housing project in Roxbury, Massachusetts, a suburb of Boston. The Bromley-Heath Housing project was among the worst in the Boston area. About 85 percent of the tenants were on public assistance; Lynda's family was no exception.

She attended Dorchester High School, which was among the poorest public school systems in the nation. The dropout rate was twice the city's average. Lynda recalls being on the verge of becoming a delinquent teenager. She states, "my friends and I would hang out and smoke and I guess maybe we were looking for a little trouble." But all that changed for Lynda when she met a concerned

Lynda M. Jordan, an associate professor of chemistry at North Carolina A & T University, rose from the Boston ghetto to the ivory towers of the Massachusetts Institute of Technology and was a featured scientist in the 1995 PBS film series Discovering Women: Jewels in a Test Tube.

teacher, Joseph Warren, in 1973. Warren was the director of Upward Bound at Brandeis University, a program for low-income urban youth aimed at increasing their self-esteem and chances for a college education. Lynda was impressed at what Joe Warren's program offered and the positive direction that he gave her. She spent two summers in the program before going to North Carolina Agricultural and Technical (A & T) University for her bachelor's degree in chemistry. She was in a predominantly black institution of higher learning where she felt accepted and valued. From there, Lynda continued graduate studies at Atlanta University in Georgia before finally going to the Massachusetts Institute of Technology for her doctorate. She was now back home in Boston—but not at home at MIT. MIT had become known for its ability to attract high-achieving minority students, but it was chilly place for Lynda. Some of her major research papers and notes were stolen. She had to repeat her experiments and learn to be more cautious and protective of her work. She faced many challenges at MIT and has been described by her professors as one of the brightest, most intelligent, and hardworking students to come through their Ph.D. program in chemistry. At that time, she was only the third black student to successfully complete the Ph.D. program in her department.

In 1985, Lynda undertook another interesting and pioneering journey when she won a fellowship to work as a postdoctoral scientist in the Institut Pasteur in Paris, France. Lynda was the first black woman to ever be accepted to this prestigious institution. Lynda worked with Françoise Russo-Marie, a leading biochemist at the institute, on PLA-2 enzyme (a protein produced during the childbirth process in the mother's placenta). This was the first time Lynda had worked one-on-one with a leading female scientist. She also loved Paris and all the cultural activities and experiences she had in France. Lynda felt that "the Europeans treated me well as a scientist. They respected me," an experience that mirrors those of other African Americans who have lived in Europe. Upon her return to the United States, she decided to work at her alma mater, North Carolina A & T University. She began as an assistant professor and progressed through the ranks.

I read about Lynda in the periodical *Black Issues in Higher Education* at my university library. From time to time, this journal profiles a black scientist. After reading the article, I knew I wanted to interview Lynda. When we met, it was like two sisters finally finding each other; because we have the same last name, we even wondered if somehow we might be related. We arranged for a phone interview at a later date.

Lynda has spent most of her career at a predominantly black institution; I, on the other hand, have spent most of my career at a predominantly white institution. Most of our experiences are similar, but they have played themselves out differently due to our locations. Lynda has raised about a million dollars to build and supply a lab with the latest equipment for some very sensitive procedures. I, on the other hand, have not had major concerns over basic research equipment, but have had other unsettling experiences. Our conversations have continued over the years and we have grown as friends, as scientists, and as people in search of their true destiny. Lynda's story represents a true example of achieving against the odds in the scientific professions. Because Lynda's story has appeared as a part of a PBS documentary series, we talked only briefly about her early childhood experiences and moved directly into the issues of race and gender for black women scientists and to the strategies that younger women must consider when embarking on a career in science. At first glance, Lynda may have appeared to be an unlikely scientist, but she is a top-notch scientist who is down-to-earth and speaks frankly about her life experiences. Her story is a true testimonial to what a little caring, nurturing, and resources to our underprivileged in the urban ghettos and poor rural areas can do given a chance.

How did you first become interested in science?

Jordan: It was from my participation in an Upward Bound Program for inner city students. My introduction to chemistry came in this program. It was the first time I even thought about it.

Was there any particular person that influenced you?

Jordan: Joseph Warren, the director of the Upward Bound Program, encouraged me to pursue science and chemistry.

What role did your family play in your selecting science as a career?

Jordan: Mother bought me puzzles and she gave me supplemental homework to make sure I was doing my schoolwork. I was a fast learner and most of my teachers in elementary school would give the homework by writing it on the board. I would generally finish the homework by the end of the day. I think the fact that my mother challenged me with puzzles probably helped my problem-solving skills.

What role did teachers or the community play in the process?

Jordan: The teachers were very good at that time in that they encouraged us and they had high expectations of us. This was very true of my elementary years. I had my first black female teacher in the fourth grade. Whatever my teachers lacked in terms of my early education, my mother made up for. In high school, that's where I noticed the low expectations of teachers came out or were really expressed. That's why the Upward Bound Program became such a pivotal point in my high school career.

> *What role did geographical location play in your selecting science as a career?*

Jordan: Well, as you know, I grew up in the Boston area and I think urban life has an effect on young people. It had no direct effect on my selection of science, but it influenced the direction of my life in the teen years.

> *Do you feel race and gender have played a particular role in how you are viewed as a scientist? If so, could you describe specifically how race has played a particular role?*

Jordan: Definitely. You have to consistently prove yourself as an intellectual being. When we walk in the door, the questions or perceptions that seem to instantly enter the mind of a person of another race are, "What are you doing here? What do you want? Who are you?" And so on. You are doing very well if the perception is that you are a teacher. Most often you are thought of as a maid or some other working-class occupation. It really doesn't dawn on people that you can be or are a scientist. That is a perception that is often held by all people, including our own race. You are always underestimated. You consistently have to prove yourself over and over and over again.

> *Yes, I can relate.*

Jordan: Too much energy is wasted on proving yourself.

> *Can you distinguish between race and gender issues?*

Jordan: They both hurt. They both feel the same and they both are painful. I can't differentiate between them because I work at a historically black college (HBCU). I guess I do have some opportunity to see the differences, but either way they both hurt. What I will say is that sexism exists in all races, but there is a distinction between racism and sexism, and they both hurt.

> *You have a different perspective from many of us. Can you cite a specific example? It's easy to talk about racism and sexism in a predominantly*

white environment, but when you are in a predominantly black environment, what are the issues?

Jordan: Well, I am trying to deal with some of those issues right now. I have a hard time speaking negatively about these HBCUs because such comments could have a major impact on the status of these colleges and universities. Society and the higher education community are always looking for ways to downgrade the HBCUs, and I do not want to do that. The validity and the contributions that HBCUs make to society are too important to compromise. But I will make a few general comments. If you would examine the number of promotions, for example, and see how many African American women are deans and chairs of science and engineering departments in HBCUs, that would be quite telling. Also, check out the teaching loads of the women, particularly the African American women, and compare that to the men. I believe you would learn a lot from that process.

Are you the only female faculty member in the Department of Chemistry at your university?

Jordan: No, there are three women in chemistry at North Carolina A & T State University, and I wasn't the first one. Still, if you examine these institutions and where the women are in the higher echelon, you will find that they are not equal to the men. The productivity of women in terms of promotion to positions is less than the men, but that's not because they are not talented. More importantly, the workload of women is overwhelming. The lack of promotion does not coincide with the work and/or the contribution made. Women are still thought to take care of everybody. Our role is looked upon as the caregiver.

I see what you mean. You are an associate professor of chemistry and you have been at A & T for eleven years. Are there other factors that have affected your life as a scientist?

Jordan: Yes. Age can sometimes be a problem. It's just so many things that we have to deal with. For instance, the issues of being a woman, and then the woman vs. woman problems. Do you want me to deal with that?

Yes, please share some of your experiences.

Jordan: There are so many barriers for young African American women to deal with. Nobody really wants to deal with us because we are seen as the lowest on the totem pole. Being an African American woman in science, we have to deal with the isolation. Once you get used to being alone, you must then deal with getting through a system that constantly invalidates you.

On the other hand, you must also deal with personal and social isolation. What I mean is that for those of us who are single, we have to deal with men who don't want to accept or deal with us. Men, particularly black men, feel insecure. So therein lies more isolation. Another issue is that we are often first-generation college graduates. Although our families may try very hard to support us, they often simply don't know how. It's not their fault, it's just lack of experience.

Let's go back to the issue of how women treat each other. That's an important issue that we need to deal with, too.

Jordan: Women have adopted the societal views of women in a leadership position. There are several articles concerning women in science that indicate that women are more comfortable supporting men as leaders, as authority figures, rather than women. What is so sad is the rejection of women by women, and that indicates where we really are in the women's movement. If we cannot support each other, how can we take our rightful place in society? It really tells us how we feel about ourselves.

Issues concerning "women of color" in leadership are even more pronounced, particularly for African American women. I do not think that nonminority women even consider us, African American women, as a part of their "club of sisterhood." We are still excluded, not only in science, but in all aspects of life and society.

Do you feel that the women's movement played any role in your selecting science as a career?

Jordan: No. The women's movement was "not down" with black women. Remember, Mary McCleod Bethune, Madame C. J. Walker, and Jewel Plummer Cobb and other unsung heroes paved the way for us. I would love to hear what they had to say now. Some of those women were not allowed to do anything else but business and education. Of course, the exception was Jewel Plummer Cobb, because she is a well-known cell biologist. There are others, too, but the numbers are so small. And for those who forged ahead, there is no record.

Do you feel that the civil rights movement had an effect on your selecting science as a career?

Jordan: The civil rights movement gave me an opportunity to be able to know that science existed. So definitely, it had an effect. Also, affirmative action was important for me.

What responsibility do we have as black women scientists in our own success?

Jordan: I am dealing with those issues now. It is very difficult for us. There is not a support system in society for us. The low expectations that exist among other scientists for us are unbelievable. I am not saying that this is the case for all scientists, because I have a collaboration with two Caucasian women and others. Their support scientifically is very helpful, and I probably would not be doing what I am doing without their support.

We have a responsibility to continue in science, despite the difficulty. A lot of black women scientists quit after 10 or 15 years of suffering, and you really can't blame them. Why not finally get rewarded with some of the things that you may want in your life by moving into an administrative position (doubled salary, own your home, travel more, etc.)? After all, these women who have gone into administration have paid their dues. This is what happens in some situations. I am not saying that you can't benefit and earn a decent wage (depending on where you are employed) as a bench scientist, but the rewards are often less and the job is time-consuming if you are trying to do research and carry a full teaching load—or sometimes a teaching overload. What is so sad is the fact that we women of African descent are looked upon as teachers. We are not seen as scientists. Not even in our own community. Therefore, our intellectual, analytical abilities and contributions are ignored and invalidated.

What can we do to be accepted in our own community?

Jordan: We have a responsibility to speak out about the issues for black women scientists and to make sure we leave the doors open for other young women coming behind us.

How have marriage and family played a role in your life as a scientist?

Jordan: They have played no role. Well, I was married earlier, but basically it has not affected my work. I don't know if that's positive or negative, and that needs to be said. It would be nice to have someone to come home to. I want to articulate that we are single not because we want to be and that some of us have had to make a choice between science and a husband, in effect. There are not many men who are truly willing to deal with a woman who may be an intellectual heavyweight. Some men have a problem dealing with someone who may be their intellectual equal/leader or is perceived as challenging in a relationship. Some sisters feel that some black women scientists are just too difficult. I tell sisters who are married or in a supportive personal relationship, "Thank God you have someone who accepts you and supports you." Lots of men are intimidated by the intellect, but they don't understand that we are humans first and not books.

So again it is a problem that we black women scientists have in the black community. You have articulated a larger issue, because we are not accepted by our community. We are perceived as a bit weird, and that translates into other isolations, such as the social and personal isolation from our sisters and the brothers that we talked about earlier.

Jordan: That's right.

Often young women view a scientist as someone who must choose between career and family. You've articulated this earlier. What advice would you offer these young women, especially black women?

Jordan: I would advise young women to have open communication in their relationships. I would also say to young women that they should marry someone who can support their career. They have to learn to balance the two (marriage and science career). The young women should make sure they nurture their relationships as they would their experiments.

Many of the women that I have spoken with have said, "Choose a mate wisely from the beginning."

Jordan: That's true, but many young women don't understand that. Women in general, and particularly African American women, need to learn how to completely love themselves before they decide to share themselves with someone else. It is important to go into a relationship knowing who you are in all aspects of your life. A relationship should accentuate, for both parties involved, the morals, ideals, goals, and commitment that each individual has already established for him or herself. I would tell young women to stop looking for someone to fill the empty spaces in their lives. The women should fill the holes themselves and make sure they are whole before joining their lives with someone.

What can we do as black women to change our image in science? What can we do to become more visible?

Jordan: One thing we can do is what you are doing. I remember when I said to you, "Let's do a book together." You said, "Lynda, why don't you do your own? There's so much to be done." So I am motivated to do my own.

It's great that you are motivated. I still would like to do a book with you. I think your story is so powerful and fascinating and it surely stands on its own merits. Furthermore, I think every women' studies department and every black studies department should have your video in their library collection and show it to their students. And that's just the start. It should also be in the K–12 curriculum. It's a power-packed story that needs to be in print as well.

Jordan: Everyone has a different angle. We need to tell our own stories and find places to publish them. None of the mainstream black magazines were interested in publishing my story when the video was released. That video on PBS was the first time in history a black woman scientist's life story was presented.

Do you think professional organizations have a particular role in promoting black women scientists?

Jordan: That's a good question. I think organizations like the American Association of University Professors can promote us from the academic side. Also, the American Association of Women in Science is another organization that can help to promote us. On a superficial level, there are people and organizations beginning to want representation, but I do not see any true concerted effort to fully embrace black women as full participants and equal partners. Non-minority women also have an issue with black women. Yes, we can be accepted now, although minimally, as singers, dancers, and actresses—but as intellectual heavyweights, no! Not yet. The stigma of the "mammy syndrome" of taking care of everyone is still how we are viewed. These stereotypical views have to be eliminated.

African American women have a responsibility. We need to be secure in who we are and we need to have the courage and integrity to stand up for ourselves and each other. Many of us are so involved in trying to seek approval from others that we assimilate to the point of losing ourselves. That, in itself, is as much of a crime as exclusion. We should support the Minority Women in Science network (MWIS) through AAAS.

Most organizations, like my own—the American Society of Agronomy—split their committees into women and minorities. I think that this sometimes really works to the black woman scientist's disadvantage because, in some ways, she is not accepted by either group. Most black women just end up going to the minority group. I think that choice results partly from the history of women in this country (particularly the relationship between black and white women) and partly from this ob-

ligation we feel for our race, that no matter what we must support our
black men. Well, that's another interview!

Jordan: Yes, that is often the case.

Tell me about your research.

Jordan: I started the research on phospholipase enzyme (PLA-2) when I
was in Paris. First of all, an enzyme is a protein just to be clear for your nonsci-
entific audience. PLA-2 is an essential enzyme that is found in every living cell
of the body. However, it is found in minute traces in human cells. Specifically,
we use the cells in the placenta to extract the enzyme PLA-20. We can go to
the hospital and get the placenta after childbirth. We use that placenta tissue
to extract the PLA-2 enzyme. There are other proteins in the placenta and it
takes a lot of meticulous work to get a pure protein. It can be a challenging and
difficult process.

Tell me about what this research would mean in terms of our everyday
lives.

Jordan: In pregnant women, the enzyme helps to trigger the uterus during
contractions or signals it during the child birthing process. If I am able to pu-
rify the enzyme and determine the structure and function, then I can start to
address some of our everyday problems. An understanding of the physiological
basis of the enzyme can help identify the underlying causes of things like hyper-
tension or kidney failure. Those are the types of issues that affect human lives.
Of course, the big challenge is getting and maintaining a pure sample to do the
studies which will be the basis for answering some of the everyday issues.

I understand that you actually were able to successfully extract that en-
zyme in human tissue when you were doing your postdoctoral training
in Paris at the Institut de Pasteur. It must have been exciting. Tell me
about it.

Jordan: In Paris, I felt great. It was a beautiful time in my life. I had received
my Ph.D. from MIT, and I was ready for the world.

I was treated very well in Europe. I worked with Françoise Russo-Marie. She was great to work with. She was the first female scientist that I had
worked with one on one. I tell you, it was great. Her lab group was working on the PLA-2 enzyme from nonhuman sources, and my question was
whether there were other PLA-2 enzymes that could be extracted from humans. I discovered that the enzyme could be found in the placenta.

It all sounds quite exciting, and it seems that you had made a significant contribution in that discovery. Could you describe how your life was as a scientist during this time?

Jordan: It felt great because I was respected for my contribution and hard work. I wasn't seen as a black woman. I was seen as a scientist. This is significant. I loved all the culture and people that I met during that time.

It was a beautiful experience.

Your experience sounds like many African Americans who went to Europe during the time of segregation. Where do you see yourself in ten years?

Jordan: Where I will be and where I want to be are two different things.

Right. So where do you want to be?

Jordan: Well, that's deep, girl. I want to do my science, but it's hard. I've been offered some lucrative jobs. People at my university want to pull you into administration, but I am not ready to go there, yet. I don't think I've made that significant contribution that I want to make in my science. I want my science, but it takes a lot of work. I want the opportunity to make a contribution that I'm pleased with and not worry about what everyone has to say about it.

I have been dealing with some of those same issues. What do you think your greatest contribution will be to science or the larger community?

Jordan: Well, I don't know what my greatest contribution will be. I think some of the work with this enzyme will be a part of it, but it has been done at tremendous sacrifice. We have so much to contribute, but there are so many needs to be addressed. There are so few of us black women scientists, so that where we decide to make a contribution will be significant.

What would you like it to be?

Jordan: I can certainly talk about the work that I've done with the enzyme, such as relationships of function and structure, etc. In fact, I know I can do that because I have achieved part of that. To go further means that I must cut some of the things I am doing and become very focused. I spend a lot of time on teaching, and that's not what I want to do. Although I believe that I have impacted some lives, I also want research productivity. I have to have it. But I would also like to bring together the spiritual and sociological aspects as they relate to our black community.

What final advice would you like to offer young women scientists, especially black women?

Jordan: Don't limit yourself to what you should do, but go deep within yourself for the answers. Follow your heart and don't let your dreams be deferred. Continuously develop yourself spiritually and emotionally. Trust in God and seek him for guidance, even in your science.

Interview Date: June 1998.

Editor's note: Since the interview, Lynda Jordan has returned to graduate school to pursue a theology degree at Harvard University. She hopes to combine a career in science and religion and use those skills to help her community.

Selected Publications and Research Activities

Aarsman, A. J., G. Mynbeek, H. Van den Bosch, B. Rothhut, B. Prieur, L. Jordan, and F. Russo-Marie. (1987). Lipocortin inhibition of extracellular and intracellular phospholipase A2 is substrate dependent, FEBS Letters, 219, 176.

H. N. Bastain, D. Wink, I. Wackett, D. Livingston, L. Jordan, J. Fox, W. H. Orme Johnson, and C.T. Walsh (1988). Hydrogenases of Methanobacterium thermoautotrophicum, strain (delta). In The Bioinorganic Chemistry of Nickel VCH Publishers, Inc, chapter 10.

Rothhut, C. Comera, B. Prier, M. Effasfa, L. Jordan, and F. Russo-Marie, (1988) A 32 KDa lipocortin isolated from human blood mononuclear cells is a calcium and phospholipid binding inhibitor of phospholipase A2, B. Journal of Cellular Biochemistry, Supplement 12E, UCLA Symposium on Molecular and Cellular Biology.

Jordan, Lynda. (1992). Purification and partial characterization of the human placental phospholipase A2 isoforms. Journal of Chromatography, 597, 299–308.

F. Radvanayi, L. Jordan, F. Russo-Marie, and C. Bon. (1989). A new sensitive fluorometic assay for phospholipase A2 in the presence of serum albumin. Anal. Biochemistry, 171(1), 103.

J. Kearneyu, S. Wolk, M. R. Schure, and L. Jordan. (1995). Conformational studies of biotinylated DNA oligonucleotides utilizing two-dimensional nuclear magnetic resonance and molecular dynamic. Analytical Biochemistry, 224, 270–278.

SHELIA MCCLURE

A Woman's Place

TWO YEARS AFTER THE FAMOUS 1954 *Brown v. Board of Education* decision created separate but equal educational facilities for African American children, Shelia McClure was born in rural Carrollton, Georgia. Being the youngest of seven children, Shelia was a joyous and curious child surrounded by a loving and supportive family. Paulding County, Georgia, was and still is the home of the McClure clan. Shelia jokingly says, "There are McClures everywhere in Paulding County." Shelia had plenty of role models in her own family to nurture her natural curiosity about nature and life. Having older siblings in mathematics and the healthcare professions exposed her to the sciences early on. Anything that was missing in terms of opportunities in Paulding County was made up by siblings and a host of successful relatives who were ready and willing to assist "baby Shelia" with any of her heart's desires.

Shelia McClure, an associate professor of biology at Spelman College and a health scientist administrator at the National Institutes of Health, finds time for research, students, and family.

It was no surprise to anyone when Shelia excelled in her high school courses and decided to go to college. An exceptionally talented young woman in many subjects, Shelia recalls being steered away from the sciences early on. She recounts, "I was steered towards a scholarship in Spanish, but I wanted to do science." This was one of her first experiences with racial biases exhibited by college recruiters and counselors.

Fortunately for Shelia, she did not allow any prejudices or misgivings by counselors to deter her from her life's goal. She completed her bachelor's degree in biology at Savannah State University in Georgia. From there, she went to the University of California, Los Angeles (UCLA) as a naive twenty-one-year-old graduate student. She was one of the first five African Americans to graduate from the zoology department with a doctorate, with an emphasis in cell biology.

She, like many other young African Americans of the 1970s, was a beneficiary of affirmative action and the long, hard struggles of the civil rights movement. Although she certainly faced challenges, Shelia excelled in her graduate studies at UCLA because she knew what her goals were, kept her eyes on the prize, and was qualified to be there. She surrounded herself with support from the community. Shelia also was introduced to her future husband during this time.

With Ph.D. in hand, Shelia decided to return to her home in Georgia. She accepted a position at Spelman College in Atlanta. Founded in 1881 for black women, Spelman College has a long history of producing outstanding black women who excel in science and other professions. Spelman can boast of being the leading producer of African American female doctorates in science. As a successful black female scientist teaching at a historically black college, Shelia is in a unique academic setting where she is surrounded by African American women who hold key positions in the college at all levels. Having progressed through the ranks to associate professor, Shelia describes her experiences at Spelman as phenomenal. She has found that her students relate to her easily and often pick up on her enthusiasm for science and the health-related careers. Shelia has been an exceptional scientist and mentor to hundreds of young aspiring African American females and has found her place at a woman's place, Spelman College, one of the truly great places for a scientist.

In addition to her duties at her home base at Spelman, Shelia has taken advantage of a wonderful opportunity to serve as a health scientist administrator at the National Institutes of Health in Bethesda. This new adventure allowed her and her family to explore new horizons.

How did you become interested in science?

McClure: I think I've always had a natural curiosity about things. When I was growing up, I was always the person who was not afraid of anything. I was always collecting insects or worms, performing "surgeries," and doing dissections. One of the things I can remember doing as a child is an autopsy on my goldfish after it had succumbed. That curiosity certainly expanded as I went to high school. My high school biology teacher was a wonderful woman who was in love with biology and really opened my eyes to careers far beyond medicine. I started to think that graduate school might be a viable option.

Was there any particular person who influenced you to consider a career in science?

McClure: I think the person who had the biggest influence on my becoming a scientist was Margaret Robinson, who was chair of the biology department at Savannah State College when I was a first-year student. Another influential person was Roy Hunter, who was a professor of physiology at Savannah State College (now University). He presented me with the chance to work in his research lab during my second year at Savannah State.

What role did your family play in your selection of science as a career?

McClure: Well, I think a lot of people in my family had an interest in science and medicine. Basically, I think it was the expectation in my family that we would do well in all subjects, including science and math. I wasn't intimidated by math and science courses. My parents were certainly very supportive of anything I really wanted to do. As a child, I can remember wanting to be president of the United States or at least having a political career. At another time, I wanted to be in the performing arts, and finally I decided to become a scientist. I also had family members interested in working in math and science areas.

Any of them siblings?

McClure: My brother is a mathematician, and my sister is in a healthcare field. I had a lot of exposure to physicians, nurses, and educators in the science and math areas. I think it was probably an unspoken expectation that I would go into science or math.

Tell me more about your K–12 teachers or the role that community played.

McClure: I honestly think my interest in science was a natural curiosity. I cannot think of any specific teachers in elementary school or middle school who really influenced me one way or the other to pursue a career in science. My teachers were very supportive of me both academically and personally. I was always involved, for example, in science projects and science fairs.

Where did you grow up?

McClure: I am a native Georgian. My father's family is originally from Carrollton, Georgia, and there are a lot of McClures in Carrollton today. I'm related to all of them! I grew up in Paulding County, a rural area in Georgia, which is about thirty minutes northwest of Atlanta. My father still resides there.

Are you the youngest?

McClure: I'm the youngest child of a wonderful family of seven. We are a very interesting family in that my niece and I are about the same age. Because of the vast differences in the ages of my siblings, I really feel that I grew up in a small family, since I was alone at home.

Do you think race or gender play a particular role in how you are viewed as a scientist? Can you cite specific examples of how race was a factor in how you are viewed?

McClure: I think both race and gender have played a role, sometimes simultaneously, sometimes individually. My decision to pursue a career in science and the actual path I took were both influenced by race and gender issues. For example, in high school I remember competing for a scholarship to college where race became an apparent issue. I was an excellent student in Spanish but I also was excellent in science. Because of my skills in Spanish, the natural assumption was that I would choose to major in Spanish. I visited the campus and discovered that all kinds of college interviews for the scholarship had been set up in the foreign language department. When I went there, I had explicitly told my career counselors that I was not interested in majoring in Spanish and that I was going to be a biology major. I thought that it would be more appropriate for me to interview for the scholarships in biology. Much to my dismay it had been assumed that Spanish would have been a more appropriate major for a black student. I wanted to see what the lab facilities were like at the college—which shall remain nameless.

The fact that people would assume that I would not choose science as a major as a first-year incoming student was disappointing. I will mention that this was a very well-known majority institution where I was interviewing for the scholarship. The ironic part is that I did win the competition for a scholarship in biology. But I opted, of course, not to attend that institution. I think a big part of my reason not to go there was the natural assumption that I would not be a science major, even though it was not explicitly stated. I think that attitude displayed by those college recruiters and career counselors really had an impact on my perception of that institution and how I would be treated as a black person at that institution.

Other experiences were from counselors at my high school who really didn't take my interest in science seriously, even though I had good grades in math and science. When I would talk with them about careers, possible majors and which colleges to attend, science never entered into the conversations. I think much of what I felt early on was racially motivated in the sense that I had counterparts who happened to be white females who were not viewed in the same manner and whose grades or SAT scores weren't nearly as good as mine. It didn't discourage me. It just said that I might want to look at an HBCU as opposed to a majority institution for my undergraduate education. So I chose to attend Savannah State College. During my undergraduate years, there was a lot of support. The chair of my department was a strong black woman who was a pioneer in her educational pursuits. She was the first African American female to receive a Ph.D. in botany from Washington University in St. Louis. As a role model and mentor, she was really the boost I needed to keep me going in the right direction.

> *Are there any particular issues or examples of gender and race that influenced you as you became a professional?*

McClure: I think that it was in graduate school when I became painfully aware that networking in the sciences tends to be very gender-biased. I did not experience any academic difficulties, but I knew I was not part of the loop, the graduate student network. No matter how hard I tried to fit in, somehow I was not part of that loop; but that was OK because I had other support built in that I utilized when I needed it. I was able to survive because I had other networks, like church, sorority, and the black community.

> *Now that you teach at a college for black women, has your experience in terms of race and gender been different from someone who has gone to a majority institution or to a profession outside of education, such as industry or government? Have you experienced race and gender biases at Spelman?*

McClure: I really don't think so, in the sense that I'm in a department where there are other women scientists. I'm in a department where there are other black women scientists and sometimes I forget how truly blessed I am in that regard. I realize it when I go to national meetings and realize there's no one in the room that looks like me. Spelman is something very special and something to be treasured. The fact is that we have four black women with Ph.D.s in the biological sciences—from Wayne State, MIT, Emory, and Berkeley. I'm so used to that that sometimes it takes other people mentioning it to make me realize that this is not the norm.

> *Are there any other factors that might affect how you're viewed as a scientist?*

McClure: The fact that we're at a small school that also happens to be an HBCU (Historically Black College and University). There are those who think that "good science" has to come from a major corporation or research university, and that is absolutely not true. While the level of scientific inquiry that goes on at a larger university may be tremendous, the quality of science need not be limited by where it takes place. Good science can occur anywhere, as long as the scientists are good at what they do and have integrity.

> *You achieved your educational goals while you were quite young. Did you experience any difficulties related to your age?*

McClure: I was very young as a graduate student. I thought everyone in graduate school was twenty-one. I found out that was not the case. I thought everyone had a Ph.D. at twenty-six. I was a bit naive in my thinking and I found that being young had some advantages and disadvantages. One of the disadvantages of being a young Ph.D. was having to prove myself to get people to take me seriously. One of the advantages is that I found it very easy to relate to my students and them to me.

> *Do you feel that the civil rights movement played a role in your selection of or success in science as a career?*

McClure: I don't know if the movement played a role in my selection of science as a career, but certainly my being able to pursue this career. I would be remiss not to acknowledge the fact that had it not been for the civil rights movement, UC-Berkeley would not have taken a risk on me.

> *Were you the first black to get a Ph.D. in your department?*

McClure: I was not the first black in my department to get a Ph.D. I was among the first five blacks to get doctorates from the Department of Zoology. In the early 80s, there were fewer than ten Ph.D.s awarded to African Americans.

How did the women's movement affect your career as a scientists?

McClure: I think that the women's movement made for a climate that was much more accepting of women. That's not to say that the climate is or was perfect. Because of the women's movement, it was much more acceptable for a woman to become a scientist and to choose a nontraditional career path.

How do you feel other scientists—particularly black men and white women—can be more supportive of black women scientists?

McClure: You would want the same level of respect for your work that any of the above groups would give to any of their colleagues. I don't know if there's anything special. Just take us seriously as scientists, as professionals, and not come through the door with certain assumptions because we're black and we're women.

What responsibility do we have as black women scientists in our own success?

McClure: I feel very strongly that we have the responsibility to train other black women scientists, as evidenced by the fact that I have chosen to come to Spelman and I am doing what I'm doing. I think we have a responsibility to provide the best-quality training to other women scientists. I think we have the responsibility to serve as mentors to other young women scientists.

Many black women scientists or black women professionals say that they already have enough on their shoulders in light of the double whammy black women face. What are some of the strategies that you would recommend to black women in terms of how they can manage this role as mentor?

McClure: I think anything you do has to come from the heart. You first have to be committed to what it is you want to do. I think that it's important that you don't let other people define for you what your role should be. Certainly there may be things I deem as important, things for me to do in my life as a scientist, and in my life as a black woman scientist in particular, that other people may not view as being very important to the scheme of things, but it becomes a matter of personal choice, I think, for everyone.

How have marriage and family played a role in your life as a scientist? Are there any strategies you use when balancing a family and science career?

McClure: My strategy is that family is first. That is my personal choice and I am happy with that choice. However, just because family is first does not mean everything else suffers in response to that. What I find is that by putting family first, I get a lot of support from my family. For example, if a progress report or grant is due, I find that my family is understanding of that and they are supportive of that. It doesn't have to be either/or: either I'll be a scientist or I'll get married. It's possible to be a scientist and be happily married and have a family. It's all about balancing and prioritizing things in your life. My husband understands that I may have to work late sometimes and he's willing to assume responsibility for our child. I'm not saying there aren't times when they have to make sacrifices that go beyond the sacrifices they would have to make if I were in a different type of career. They're willing to do that because they know that in the end my devotion lies with my family. By the same token, I'm not going to shirk other responsibilities. I'm going to perform my job in a professional manner. I'm going to be involved and be a contributing member of the scientific community.

> I'm sure a lot of young women want to hear that. Because of the long hours and rigor in science and engineering, often young women view being a scientist as having to choose between career and family and personal life. Is there advice you'd offer a young woman, particularly a young black woman?

McClure: Again, it is about personal choices. One can choose to be married, and that's okay. One can choose not to be married, and that's okay, too. You should not feel like you have to fit into a particular mold or choose a life that someone else has decided your life should look like. You decide what your life should look like and choose a career that you're genuinely interested in and that you love so that the work does not seem like work. It's more of an extension of self. For me, my husband is and has to be the kind of person who understands that childrearing takes the both of us. He knows it's okay to change diapers, it's okay to go to PTA, it's okay to do those "female things." Those things don't have "For Women Only" stamped on them. For those women who are involved in relationships and want to have a science career, they have to have partners willing to cooperate to support her career.

> That's wonderful advice. What can we as black women do to become more visible in science?

McClure: That's a good question, too. I think we have to become more involved in national and international initiatives. I think we have to participate on research panels where we can truly have some influence. If there are not venues where our voices can be heard, I think we have to create venues. I think that black women are creative and resourceful. We need to enlist some of that creativity and resourcefulness in trying to start initiatives where we do things more formally than informally, like mentoring programs. I think we already do much of that, but we need to do more at a higher level to gain more national and international exposure.

> *Do you think professional organizations have a particular role in promoting the visibility of black women and, if so, what do you think that role should be?*

McClure: I think professional organizations can increase the visibility of all women in general, and in particular black women. I think many societies have started to address these concerns. I think of Women in Cell Biology, for example, as being an important part of the American Society for Cell Biology. I know that a lot of the national organizations have women's groups. Again, we need to be sure we are putting issues that need to be addressed on the table and identifying areas that are unique to black women scientists.

> *Like what?*

McClure: The sponsoring of more panels that look at these kinds of issues that we have discussed. Tracking and finding out what happens to young black women in science, how they choose different careers and how they're supported or how they feel they have been supported in these careers. Research which looks at black women scientists at HBCUs as opposed to the majority institutions. Those types of issues for black women scientists.

> *What do you think will be your greatest contribution to science?*

McClure: I think my greatest contribution will be that I have addressed a problem that affects African American women and hopefully some of the findings will shed light on possible treatments for uterine tumors, which affects a disproportionate number of African American women.

> *Is that your primary research? Tell me a little more about it.*

McClure: In our lab I have developed primary uterine leiomyoma cell lines. We utilize these cell lines to study factors which promote their growth *in vivo*. We are trying to gain a better understanding of why these particular agents promote the growth of these tumors. In this work we have also been able to identify a growth-stimulating factor. We hope to resolve the identity of this factor.

How long have you been doing this research?

McClure: This has been essentially my life's work since I've been here at Spelman. I think it's been about twelve years, starting from developing the cell lines, characterizing the cell lines, and monitoring their growth response to the point where we have identified this growth stimulatory factor.

And what will this work mean in terms of black women's health?

McClure: Hopefully, it will shed some light on the etiology of these tumors, how we can treat them without surgical intervention and—dare I say?—how we might be able to prevent them or at least predict when they might occur.

Is there anything else you'd like to add to that contribution in terms of sciences?

McClure: That I had a great time doing it.

You train a tremendous number of black women at Spelman. This has to be a major contribution to science. Spelman College has trained many black female scientists, and you're a big part of that.

McClure: That's true. I hope that students have been able to appreciate the fact that I am here, because I want them to benefit from my experiences. Furthermore, I hope they have seen the enthusiasm and love that I have for science and that this will also attract them to careers in science.

Nicely stated. To give readers some feel for the impact of your biology department and Spelman's science program, how many biology majors are here?

McClure: We have approximately 250 majors.

That's impressive. Finally, what advice would you offer young women, especially young black women?

McClure: Have confidence in yourself. Be good to yourself. Know that you deserve the best and expect the best.

Interview Date: October 1997.

ETHELEEN McGINNIS-HILL

It's a Good Thing

ETHELEEN McGINNIS WAS BORN SHORTLY AFTER World War II in Birmingham, Alabama. She was one of six siblings and the first to graduate from college. In the heat of the civil rights movement in the 1950s and 1960s, Birmingham was a city in much turmoil, especially after the bombing of the Sixteenth Street Baptist Church, which killed four young black girls. Because of the movement and Dr. Martin Luther King, Jr., Etheleen was inspired to achieve something special in her life. Although her parents were not formally educated, they encouraged Etheleen and her siblings to pursue higher educa-

Etheleen McGinnis-Hill, an associate professor of microbiology, finds her peace and scientific comfort at Meharry Medical College.

tion. After high school graduation, Etheleen went to Knoxville College, a small, predominantly black college founded in 1875 as part of the mission of the United Presbyterian Church of North America. At Knoxville College, she was fascinated by her biology courses, especially microbiology and cell biology. They were her favorite subjects, and she was soon hooked on a career in science.

From Knoxville College, Etheleen journeyed to Purdue University to study for her master's and doctoral degrees under the tutelage of Luther S. Williams. She felt she had a unique experience in working with one of the few black professors at Purdue University at that time. She had a very positive experience at Purdue and remembers it fondly.

After finishing her doctorate in 1976, Etheleen returned home to Birmingham, where she did postgraduate training at the Laboratory of Molecular Biology at the Birmingham Medical Center of the University of Alabama. She worked there for three years before joining the faculty at Meharry Medical College in Nashville. In 1980, Etheleen became the first black woman faculty hired in her department.

Etheleen has had a variety of research and service experiences. She has received support from the National Science Foundation, the National Institutes of Health, and several other agencies. She has also served on the Minority Access to Research Careers (MARC) Review Subcommittee. It wasn't easy for her to forge out a successful scientific career as a single parent. She says, "In the early days, I was a single parent so I had to make sure my son was taken care of first and find ways to support my research goals." Eventually she found that having collaborators was a good strategy for her success as a scientist. Although her son is now an adult, she continues to collaborate with scientists within and outside of the college. In 1986, she was a visiting professor at Johns Hopkins University in the biology department. This experience allowed her to build further collaborations with other scientists around the country.

Meharry offers a unique environment for training and educating primarily African American students in the biomedical sciences and the health professions. To that end, Etheleen has been involved in teaching and coordinating the department's core courses as well as the dental and medical microbiology courses for students in the School of Medicine and Dentistry. Because Meharry is still a small, historically black institution and Etheleen is only one of a handful of African American female faculty members, she finds that there is no shortage of service responsibilities, duties, and demands on her schedule. Despite her limited time, she is determined to make sure the next generation of biomedical researchers is properly trained and exposed to meaningful research experiences. She is active at Meharry and in the community at large in service and civic organizations. She states, "there are a lot of ways to significantly contribute to educating the next generation." For now, Etheleen is comfortable with her role as researcher, teacher, and mentor. Being at Meharry with all of its uniqueness has been a "good thing" for this native Alabamian.

How did you first become interested in science?

McGinnis-Hill: I fell in love with science in high school when I took my first biology course. I was a biology major at Knoxville College, and I fell in love with microbiology and cell biology. It was in my junior year that I decided I would go into research.

Was there any particular person who influenced or encouraged you in science?

McGinnis-Hill: I guess my high school biology teacher. She was impressed, I think, with my enthusiasm. I was doing well in her course and I was one of the top students.

Was your teacher an African American?

McGinnis-Hill: Yes, she was.

Did your family play a role in your selection of science as a career?

McGinnis-Hill: Not much, because I was the first person to graduate from college in my family and they just encouraged me to pursue higher education. They didn't care what I majored in and they supported me wholeheartedly, one hundred percent.

Where did you grow up? Do you think your geographical location influenced your selection of or interest in science?

McGinnis-Hill: I grew up in Birmingham, Alabama, and to some extent growing up in that area had an impact on me in a number of ways. Growing up there in the South did not particularly influence my decision to select science; however, I came along around the time of the civil rights movement, Martin Luther King, Jr., and segregated schools. That era influenced me to do something with my life and to be an achiever.

How many siblings do you have and are they in science?

McGinnis-Hill: I have two brothers and one is deceased. I have two sisters. One sister graduated from Alabama State and my other sister attended Alabama A & M College (now University). My brothers didn't attend college. I did my graduate work at Purdue University in Indiana and I received both my master's and Ph.D. degrees in biological sciences with an emphasis in microbiology.

Do you feel race or gender had a particular role in how you're viewed as a scientist?

McGinnis-Hill: I think race has played a role. In graduate school, however, race played a minor role because I was in a semi-protected environment. My preceptor was black at Purdue and we had other black students in the laboratory, so I was surrounded by other black people. I attended classes and did everything else with those same students. In effect, I had a support system. In some ways, I felt shielded from some of the usual issues with race. Most of my professors were white, but I did not feel discriminated against. I shared this with another individual, who was surprised. If anything, I felt I was given the benefit of the doubt.

That's great. We need to hear those stories. When were you a graduate student at Purdue University?

McGinnis-Hill: I was a graduate student from 1971 to 1976. I would like to share one example of an experience that I had while I was a graduate student. During my first year I was enrolled in a microbiology course. At the end of the course, I think my average was a B+. The professor said to me, "You know you've done well in this course" (it was a difficult course) "and you're just a few points from an A. Would you mind coming to my office and answering a few questions orally so I can get a better feel for how well you understand the material?" He also said, "You don't have to do it. It's up to you." So I said, "Sure, I'll come." I did that and we talked about different kinds of things. I didn't know what was going to happen, but I got an A in the class.

So that's just one example. In general, the professors that I took courses from were supportive. I didn't feel discriminated against at all. I think people can look beyond race and gender, in many instances, if they see that you're serious and you want to succeed. I think people will respect you for that. I'm not saying that it will be true for everyone, but that was my experience.

So for you, neither race nor gender affected how you were

viewed as you were training for your science career. What happened as you began your career?

McGinnis-Hill: To some extent, race and gender have affected how I am viewed as a scientist, but I did not start to feel it until I was a postdoc and later as a faculty member at Meharry.

Do you have tenure?

McGinnis-Hill: At Meharry, there has been a moratorium on tenure for several years. There are not many people who have tenure around here.

That is a very different academic policy.

McGinnis-Hill: It's different. In most places, once you become an associate professor, you acquire tenure. That's not true here. We have many associate professors but they're not tenured.

You obviously do not feel any threat of losing your job.

McGinnis-Hill: No, not really. I just haven't worried about it.

Are you the only African American female in this department?

McGinnis-Hill: I am the only African American female in the department. There are three other women, and they are Caucasian.

Are you the first?

McGinnis-Hill: I am the first. This department was recently merged with the Division of Biomedical Sciences, so now the Department of Microbiology has become very large. There was a black woman in the former Biomedical Sciences Division, and she left at least five years ago. That whole unit was all Caucasian except with a black chief. Of course, we're under one chief and she's Jewish, Shirley Russell. In this combined department, there are really only three blacks: myself, Robert Holt, and George Hill.

How many black women do you have on the faculty here at Meharry?

McGinnis-Hill: There are only four black women in the basic sciences: that's myself, Dolores Shockley in pharmacology, Evangeline Motley in physiology, and Marilyn Thompson in biochemistry.

I expected to find a lot of black women here!

McGinnis-Hill: You did? Bless your heart!

*This is just amazing! Do you think that the civil rights move-
ment played a role in your success in science as a career?*

McGinnis-Hill: The civil rights movement played a role in my success as a
scientist. I was inspired by the movement and Dr. King.

*Did you participate in any of the civil rights activities when
you were growing up in Birmingham?*

McGinnis-Hill: I was an onlooker. I didn't participate in the marches because I was just that focused on school. So when the kids were marching and
they all went off to jail, I was not in the group. I was in school. My girlfriend
and I were the only ones in class one day because all the other students were
either marching, in jail, or at home. I figured that it was just as important for
me to go to school, to achieve, and do well. That was my way of contributing
to the advancement of my people, because the way I looked at it, the world
recognizes intelligence and achievement. I don't think you can make people
respect you. If you want to get true, meaningful respect from anyone, then
you have to carry yourself in a certain way. You have to be competent at what
you do. You have to have something that the world values and can respect.
We, as black people, want to get the world's respect, but we have to be clear
about how we are getting it.

*Do you feel that the women's movement played a role in your
selection of and success in science as a career?*

McGinnis-Hill: No, because I had made that decision before that movement took off.

*How do you feel other scientists—especially black men and
white women—can be more supportive of black women scientists? What is the responsibility of the black woman in her
own success?*

McGinnis-Hill: I think the responsibility of other scientists would involve
their being more sensitive to the plight of the black woman scientist and her
predicament and position relative to others in the area. In general, be willing
to assist her. By the same token, black women's responsibility is to show more
aggression in seeking the assistance we need from the scientific community.
In general, you have to not be shy about asking for help. We need to realize
that we can't make it by ourselves. For that matter, no one can. We have got

to be willing to reach out, find the help, and have the right attitude to get what we need. It's not going to be easy, but it's a matter of identifying the appropriate people that are willing to assist us.

> *How have marriage and family played a role in your life as a scientist?*

McGinnis-Hill: I've been married seven years, and it hasn't played a major role. But let me say, I was a single parent for a long time and that was a challenge. I had to make a choice, in many instances, between my job and going home to do things for my child. In that regard, I would say my career was affected because of it. I was not always able to get things done in a timely fashion because I did have to make that decision. My career was influenced to some extent by my being a single parent.

> *Are there any particular strategies you recommend in balancing a family and a science career?*

McGinnis-Hill: You need to work and align yourself with other individuals who can help minimize the load that you carry as a family person and scientist. In terms of your research and other academic responsibilities, there's no way you can do it all by yourself. Even if that were not the case, the fast pace of science requires a support system.

> *So are you recommending an interdisciplinary approach to research?*

McGinnis-Hill: Collaboration is the way to go. That's the only way you can get things done in a timely fashion. I guess that's more easily said than done, but that's something you have to do. If you can't form them within your institution, sometimes you have to go outside your institution and develop collaborative arrangements.

> *Because of the long hours and rigor in science and engineering, often young women view being a scientist as having to choose between career, family life, and personal life. Is there any advice you would offer a young woman, especially a young black woman?*

McGinnis-Hill: You know some people say, and some women firmly say, you can do it all. Perhaps that is true. But I don't know how well you can do all of it. It's very difficult to do all those things well. Something is going to suffer. You're going to find yourself making a choice—either your children over the job or the job over your children. I think we're deceiving ourselves if we continue to say that we can do it all, and I don't think we can do it all equally well. That being the case, I think if you insist on doing all these things, then you need to have adequate support systems.

> *What can black women do to change their invisibility in science? How do we become more visible?*

McGinnis-Hill: We have to become more aggressive and align ourselves with groups of individuals who can assist us in being more productive. I think that's the key—to recognize that we make our opportunities and stop just sitting back waiting for someone to see us. We need to make ourselves seen.

> *Do you think professional organizations have a particular role in promoting our visibility? You seem to be saying that we should be responsible for that role.*

McGinnis-Hill: In other words, we can make it happen instead of just fading into the woodwork. We need to be more aggressive about what we're doing.

> *How do envision yourself as a scientist? Where do you see yourself in ten years?*

McGinnis-Hill: I probably won't be doing benchwork anymore. I'll probably be moving into administration. I've been involved with many student committees on campus and I'm a student advocate. I like to work on behalf of students and I can see myself moving in that direction. Right now, I have two graduate students in the lab and I just graduated one. By the time these last two get finished, I probably will move into administration. I'm trying to decide whether or not to take on another student. That would mean staying here in the research lab exclusively for another five years.

> *That's a lot of energy to educate and train students and keep a viable research program.*

McGinnis-Hill: It takes a lot of energy, and you have to have money to support the lab and the students for the duration of their time here. It's a hard job, so I don't know whether I want to continue with it that long. I think I can do some other kinds of things having to do with minority students that are just as beneficial.

Do you have any master's students?

McGinnis-Hill: No. I had one master's student in the past, but we really don't train at the master's level.

Would you tell me a little bit more about your research?

McGinnis-Hill: The overall objective of our research is to elucidate the molecular mechanisms of binding and entry that are involved in the invasion of host cells by the hemotrophic bacterium *Bartonella bacilliformis*. (*Bartonella bacilliformis* is a bacterium that can affect the human red blood cells.) The intracellular location of this organism results in the manifestation of two distinct clinical syndromes—known as Oroya fever and verruga peruana—owing to the predilection of the organism for human erythrocytes and endothelial cells, respectively. The Oroya fever syndrome is characterized by the presence of the organism on the inside of erythrocytes and the verruga peruana phase of the disease in the tissue, chronic phase in which hemangioma-like lesions can be found on the face and lower extremities. It is hypothesized that the organism binds to receptors on target cells via specific adhesions which stimulate the formation of a linkage between the membrane and cytoskeleton that promotes the internalization of Bartonella into cells. Thus, most of our work has been focused on the identification of important bacterial ligand/receptor interactions between the organisms and host cells in addition to characterizing signaling phenomena that may accompany such interactions.

How long has this project been going on?

McGinnis-Hill: Since the late eighties. I have another graduate student who's focusing on the identification of the Bartonella bacilliformis receptors in erythrocyte.

What do you think your other contributions to science will be?

McGinnis-Hill: I think that's where I've made a good contribution, as well as training black students and trying to get money for black students. I serve on the Minority Access for Research Careers committee as a reviewer. I think that helps, as we can keep the money coming into our minority schools. That's what I want to do when I give this up—to continue working in that arena.

Do you only have graduate students? What about technical support for your research?

McGinnis-Hill: Off and on, but for the most part none. There's a problem sometimes in finding good technical support, even when you have the money. Unfortunately, if you're at Meharry, the problem has been maintaining good technical support. Sometimes you can identify a good technician, but they don't stay very long. They leave and go over to Vanderbilt for higher salaries. I'm overwhelmed sometimes by all the tasks that I must do in order to keep going, in addition to teaching responsibilities and committee assignments.

How many classes do you teach?

McGinnis-Hill: I teach the medical microbiology course and I coordinate physiology and genetics sections in that course. I teach dental microbiology and a graduate course in microbial physiology and microbial genetics. The second semester we're involved in graduate course teaching in either physiology or genetics. The first semester we're involved in the medical microbiology class, and sometimes we also have to teach the laboratory component of the course. On top of all of that, I have several committee assignments.

What advice would you offer young women, especially young black women?

McGinnis-Hill: I would say work hard and become as good as you can be in whatever area you decide to pursue. Be more aggressive about what you want. Identify and seek out your own opportunities. Don't expect people to come to you and recognize you, because really no one owes you anything. You have to make your own successes.

When I got my first grant to work in this particular area, I needed someone who was an expert in blood to serve as a mentor. I don't know if you are familiar with these Minority Faculty Development Awards. They are five-year awards and they pay just about one hundred percent of your salary, but there was no one here who could serve as mentor. One of the stipulations is that the mentor must be at a majority school, and they have to be expert in whatever area you're proposing to do your work. So I just picked up the phone, looked in the Vanderbilt University phone directory for someone who specializes in blood research, and I just started calling. I identified myself and told my prospective mentor what I was trying to do. I lucked out and found a very nice person who was very good in his area and nationally known who agreed to serve as my mentor. This is what I mean by helping to make your own opportunities and not being too intimidated to do that kind of thing. Although they can say no, I guess that's about the worst thing that can happen. Sometimes you just have to get out there, and we need to do more of that. I think that's part of the solution.

Selected Publications and Research Activities

McGinnis, E., and L. S. Willliams. 1971. Regulation of synthesis of amino-acyl-tRNA synthetases for the branched-chain amino acids in *E. coli*. J. Bacteriology, 108, 254–262.

McGinnis, E., A. M. Sarrif, and K. L. Yielding. 1983. Involvement of deoxyribonuclease activity in the differential sedimentation rates of nucleoids from non-transformed and transformed mouse embryo fibroblasts. Mechanisms of Aging and Development, 22, 219–232.

Parham, C., E. Cunningham, and E. McGinnis. 1988. Differential effects of DNA gyrase inhibitors on the genetic transformation of *Neisseria gonorrhoeae*. Antimicrobial Agents and Chemotherapy, 32,1788–1792.

Hill, E. McGinnis, A. Raji, M. S. Valenzuela, F. Garcia, and R. Hoover. 1992. Adhesion to and invasion of cultured cells by *Bartonella bacilliformis*. Infect. Immun., 60, 4051–4058.

Williams-Bouyer, N. M., and E. McGinnis-Hill. 1999. Involvement of host cell tyrosine phosphorylation in the invasion of Hep-2 cells by *Bartonella bacilliformis*. FEMS Microbiol. Letters, 171, 191–201.

Buckles, E., and E. McGinnis. 2000. Interaction of *Bartonella bacilliformis* with human erythrocytes. Microbial Patho., 29(3): 165–174.

Research Support

National Science Foundation Minority Research Grant #2 S06 RR08037: Role of DNA superstructures in the control of cell proliferation: Genetic transformation in *Neisseria gonorrhoeae*, $240,000.

National Science Foundation/Minority Research Centers of Excellence; Subproject: Cloning of a genetic locus encoding the invasion phenotype in *Bartonella bacilliformis*, $8,300.

National Heart, Lung, and Blood Institute Minority School Faculty Development Award, Cloning of the invasiveness gene from *Bartonella bacilliformis*, $278,494.

National Cancer Institute, Kaposi's sarcoma is an infection-related sarcoma, $68,293.

National Institutes of Health, Minority Biomedical Research, Molecular mechanisms of *Bartonella bacilliformis* invasion of host cells, $83,692.

JENNIE R. PATRICK

Rebel with a Cause

COURAGE, CONVICTION, AND COMPASSION ARE THREE WORDS that describe the heart and soul of Jennie Patrick. These three characteristics were instilled in her at an early age from loving and supportive parents. Jennie grew up in Gadsden, Alabama, a small town in northeastern Alabama, during the segregated era of the 1950s and during the civil rights movement. In fact, Jennie was one of the first African Americans to integrate her high school. Jennie describes the experience in a 1989 article for journal *Sage:*

> The emotional, psychological, mental, and physical violence against us was difficult to comprehend. This experience opened up a whole new world for me. Surviving at Gadsden High, the previously all white

Jennie R. Patrick, a chemical engineer, scholar, and consultant, was the first black woman to earn a doctorate in chemical engineering in the United States and the first 3M Scholar at Tuskegee University.

school, became my greatest challenge. Not only did I survive academically, but also emotionally, psychologically and physically.[1]

Jennie did better than just survive at Gadsden High; she enrolled at Tuskegee Institute (now University) as a chemistry major, later becoming the first student to enroll in the newly formed chemical engineering program. Unfortunately for Jennie, she did not receive encouragement and support from the department chairman. Undaunted in her quest for a quality education, Jennie regrouped and moved west to the University of California at Berkeley. Times were financially difficult because Berkeley did not provide scholarships for transfer students. Being committed to her goals, Jennie worked initially at odd jobs; once again, she survived, graduating with a bachelor's degree in chemical engineering in 1973.

Jennie headed back east to continue graduate studies at MIT. She describes her days there as sometimes very intense, but mostly positive. As always, Jennie Patrick was ready for any challenge. Patrick became known around MIT circles as a tough, strong black woman who knew what she wanted—a rebel with a cause. That reputation, of being an advocate and fighter for students, women, and African American rights, still lives on among other MIT graduates. Not only did she leave her legacy of leadership; in 1979 she also became the first black woman to receive a doctorate in chemical engineering in the United States.

Armed with her Ph.D. and a fighting spirit, Jennie embarked on an exciting career. She began as a staff engineer at General Electric Research in Schenectady, New York. It was an exciting career opportunity where she developed a research chemical engineering program in the area of supercritical fluid extraction technology. Because so little research had been done in this area, it left a world of opportunities open for Jennie. While heading the supercritical extraction program at the Phillip Morris Company in Richmond and later holding a managerial position at Rohm and Haas Research Laboratories in Bristol, Pennsylvania, she excelled in her profession, and achieved a satisfying personal life as well. She teasingly says, "Loneliness has never been an issue for me. I have and have had a delightful personal life." Indeed, she has. Her husband, a fellow engineer, is now a successful medical doctor.

In 1990, Jennie and her husband moved back to Alabama to care for her parents. Like so many other African American parents, Jennie's were

---

1. Jennie R. Patrick, "Trials, Tribulations, Triumphs," *Sage, A Scholarly Journal on Black Women* 6 (1989): 51–53.

exceptional in their support for their children. Jennie saw this as a time to give something back to them. As she achieved this goal, moving back home led her to another exciting career opportunity, becoming the first 3M Eminent Professor/Scholar at Tuskegee University. In conversations and in her motivational writings, Jennie talks about the importance of teaching young African Americans strategies for survival in hostile environments. Tuskegee provided this opportunity in a very real way. At Tuskegee, her students observed and appreciated the talent she brought as a highly skilled engineer, a gifted teacher, and, most importantly, an advocate for their survival in a country that was still racially insensitive. Jennie has overcome personal loss and external challenges that have helped her become a person of strong courage, conviction, and compassion. She continues her fight for justice as a consultant and writer in Texas. She may claim to be "just little ole Jennie," but she is also an outstanding engineer, leader, teacher, and, most importantly, she is a woman of great character.

Did your parents influence you to become a scientist?

Patrick: My parents had very little formal education. Consequently, they did not have a major influence on the educational path I selected. However, the fact that they provided a home filled with love, discipline, and encouragement gave me the necessary foundation to achieve.

Also, I had both a curiosity and a desire to understand what and how things worked. Living in a rural community in Gadsden, Alabama, nurtured my fascination for nature. I was always filled with questions and wonder about all sorts of things.

Was there a particular teacher who influenced you to become an engineer?

Patrick: Not really. However, several of my elementary school teachers positively influenced me. Initially, I didn't understand what an engineer did, but I always heard people speak highly of engineers. Coming out of grammar school, I knew I wanted to do something in the sciences. Yet it never occurred to me that I would become an engineer.

But you knew that science was a career possibility?

Patrick: Yes, I did, because it was clear to me that science answered many of the questions I had.

Where did you attend college?

Patrick: I started my undergraduate education at Tuskegee Institute (now Tuskegee University). In fact, I was the first person to sign up to study chemical engineering at Tuskegee. The first chemical engineering department at Tuskegee didn't survive. So I transferred to U.C. Berkeley to complete my bachelor's degree in chemical engineering. From Berkeley I enrolled in Massachusetts Institute of Technology (MIT) and earned my doctorate in chemical engineering.

How did you know that chemical engineering would be a good major for you? Did you research the program at Tuskegee?

Patrick: I didn't know, especially since I wasn't totally sure what chemical engineers did. But since I enjoyed chemistry and did very well with the subject matter, I felt that the combination of chemistry and engineering probably would be more of a challenge for me. Also, I had a friend who encouraged me to consider chemical engineering. I majored in chemistry my freshman year and changed my major to chemical engineering during my sophomore year as the new department opened at Tuskegee.

When were you there?

Patrick: I started at Tuskegee in 1967. Then I transferred to U.C. Berkeley in 1970, and obtained my bachelor's degree there in 1973. I earned my doctorate from MIT in 1979.

In looking back, what were your experiences as a female pursuing a career in science?

Patrick: When I attended Tuskegee, there were very few females pursuing degrees in engineering. But what particularly struck me was the reaction of the department head of chemical engineering, who happened to be a Caucasian, male, and a Southerner. He was very negative, terribly unsupportive about my desire to become a chemical engineer. In fact, I received no encouragement from either the dean of engineering or the head of chemical engineering at Tuskegee in my pursuit to study engineering.

So when you moved on to Berkeley, were things better?

Patrick: No, at Berkeley, there were no African American female undergraduates in chemical engineering. However, there were two Asian female chemical engineering students. During my second year there, an African American male enrolled in the master's degree program. I was told that Berkeley had not had an African American in ten years in chemical engineering prior to my coming. The environment in engineering at Berkeley in the early 70s was extremely racist as well as sexist. For example, in my senior chemical engineering design class the professor forced me to do my design project alone. However, the others students worked in teams of four persons.

What was your experience at MIT?

Patrick: MIT was a different kind of experience. Since MIT is primarily a technological school, professors and staff were more accustomed to seeing Africans and African Americans studying engineering and science. There were more black people studying at MIT than I had previously seen in technical fields. In chemical engineering, there were perhaps four or five black graduate students, including African graduate students. Also, there was a comparatively impressive number of female students.

After leaving MIT, how did race and gender affect your professional experience?

Patrick: After graduating from MIT, I joined the General Electric Research Center in Schenectady, New York. My experience at GE was positive. GE at that time was an extremely competitive and prestigious research environment. Of course, there was some racism and sexism, but neither appeared to have caused me any real problem or threat at that time.

Has race or gender been a factor or concern to you in your career?

Patrick: I believe that my race has been more of an issue career-wise. It appears that many individuals tend to first perceive you as a black person. Then they recognize on second thought that you're a female. I have often observed initial reactions of intimidation or rejection of my blackness to be so strong that any secondary response to my gender is minimized.

What impact did the civil rights movement or the women's movement have on your career or selection of science as a career?

Patrick: I believe that both the civil rights and the women's movements had a positive impact on my career. Both movements increased the awareness of some of the issues which African Americans, other people of color, and women face. However, neither movement influenced my career choice.

What happened once you left GE?

Patrick: After I left GE, I joined Phillip Morris Research Center in Richmond, Virginia. I stayed there a couple of years. Then I joined Rohm and

Haas Research Center in Bristol, Pennsylvania, for five years. From Bristol, I moved to Birmingham, Alabama, for family reasons. My elderly parents were in failing health at the time and I decided to return to Alabama and take care of them. I had always felt an enormous commitment and responsibility to my parents. I wanted to provide them in their remaining days the best life and care possible. Even though I preferred living in other areas of the country rather than Alabama, my parents' welfare was my first priority.

Tell me about your position at Tuskegee.

Patrick: I presently hold the 3M Eminent Scholar Professorship in chemical engineering. This endowed chair is sponsored by the state of Alabama and the 3M Company, headquartered in St. Paul, Minnesota. In this position, I teach as well as do research in material science.

Describe your experience at Tuskegee.

Patrick: My experience here has been fulfilling and rewarding. Being here provides me the opportunity to make a difference in the lives of students. I've been able to teach them and share many of my experiences with them—thus providing them hopefully with some insight into what the future may hold for them.

I am aware that you have what is known as an environmental illness. Do you feel comfortable sharing information about your illness?

Patrick: Yes, while working in the chemical industry I was exposed to hazardous and toxic chemicals. The chemicals from the laboratories in the building were vented outside to the roof. Unfortunately, for a variety of reasons the building's intake air vent sucked the chemicals back into the building and distributed the chemical through the ventilation system throughout the building. This exposed the building's occupants to generous amounts of toxic and hazardous chemicals.

How has this exposure impacted your life?

Patrick: This exposure totally altered my life and career in the most devastating ways possible. It has limited my life in ways most people can't imagine. Massive chemical exposures most often destroy the immune system and adversely affect most major organs and the respiratory system. To survive this very debilitating environmental illness (sometimes referred to as multiple chemical sensitivity), I've had to be very assertive, creative, determined, and very positive about overcoming the illness's challenges.

I've never felt sorry for myself or wanted others to feel sorry for me. As I have said, I've taken it simply as my challenge to interact with others as normally as possible. As a professor, I've never missed a class or lecture. It's been an interesting experience because I've had to rely on others' sensitivity and openness. For instance, I've had to request that students not wear scented personal hygiene products to the classroom. Many of them respected the request immediately; others did not and presented quite a physical and emotional challenge for my survival. Usually it was the student who was least interested in learning that posed the biggest problem. However, peer pressure has been remarkably successful in assisting me. After a while the students developed a deep respect for me because they understood what I had and wanted to offer them. Also, they started to respect my deep commitment to them. Thus, they realized that it was in their best interest to be sensitive to my health issues and needs.

> I don't think that many African Americans or underprivileged groups are as aware as they should be of environmental issues, especially those related to hazardous chemicals and the impact of water quality, improper use of pesticides, etc. in their neighborhoods. Although much of the environmental justice movement is at a grassroots level, environmental literacy is still very low. Because you have to live with a condition that has been caused by hazardous chemicals, what are your feelings on this issue? Do you have any specific comments about how we can become educated in this area?

Patrick: Environmental dangers are all around us. A person does not have to be exposed to industrial hazardous and toxic chemicals to become seriously environmentally ill. Many personal hygiene products, such as perfume, hair spray and gels, and hair perms and dyes can cause serious environmental illnesses such as mine or other major health problems. The acrylic fingernail products can be quite harmful. People use these personal hygiene products and household cleaning products without thinking about the potential harmful impact on their bodies and health.

Amazingly enough, through marketing tactics many people feel they must use sweetly scented products in order to feel clean. Unfortunately the majority of these products are made from very harsh or toxic chemicals. Small children are particularly susceptible to harmful effects from such products. The African American community has a very high incidence of death and disease related to asthma. Yet African Americans from a variety of backgrounds go to physicians and expect answers and miracles without evaluating what habits or lifestyles may contribute to their illnesses or health problems. African Americans need to become more thoughtful about the products that we buy. Also, we need to evaluate more carefully the products that are marketed to us. African American communities in the past and perhaps the present have been used as the dumping grounds for industrial waste. As a whole, we need to become more observant in our communities of health patterns and issues. Also, we need to utilize the conveniences of the computer to research ingredients in the products we use.

> *That's very true, and I appreciate your sharing your story so that more Americans are aware of the environmental health issues. I would like to shift gears to discuss African American females in engineering. Are there many African American female members in the engineering school?*

Patrick: No, there is one other female faculty member in the engineering school, but she is not African American.

> *Do you have a rough estimate of African American students, particularly female, that are in the College of Engineering?*

Patrick: I am not sure about the female percentage for the whole school of engineering, but for chemical engineering the female population is about 60 percent. There are between 130 to 140 students in the entire chemical engineering department.

> *What professional organizations do you participate in? Do you feel they serve your needs as an African American female?*

Patrick: To be honest, these days I'm not very active in organizations. Due to my environmental illness, I cannot tolerate large crowds. I am a member of the National Society of Black Chemists and Chemical Engineers and the American Institute of Chemical Engineering. I still attempt to attend a few professional meetings and conferences.

In the time I have available, I choose to interact with young African Americans professionals and students. Based on my personal experiences and insight I try to teach them survival skills. I emphasize the importance of recognizing and acknowledging the true nature of one's environment. Whether or not an environment is hostile, stressful, racist, supportive, or challenging, an individual cannot be very effective or successful without honestly under-

standing and acknowledging the reality of that environment. Equally important, one must recognize that even though the playing field may appear to be
the same for everyone, in actuality it isn't.

> *You are the first African American woman to receive a Ph.D. in
> chemical engineering, so I can see how your time and schedule gets
> stretched.*

Patrick: Yes, it is important for me to prioritize and decide how to most
effectively utilize my limited time in order to communicate with as many
people as possible.

> *I noted that you have done research at several universities and with in-
> dustry. Please tell me briefly about some of your research in the past.*

Patrick: Without going into detail my area of expertise is in thermodynamics. I have used the basic concepts of thermodynamics to address practical engineering problems. My main focus of research over the years has
been in the area of supercritical fluid extraction technology. Supercritical
fluid technology uses a solvent above its critical temperature and pressure.
In doing so, some interesting and unique thermodynamic phenomena occur. The selected solvent becomes a super solvent resulting in some unique
process capabilities. This technology is primarily a separation and purification technology used in a variety of industries, such as the food, tobacco,
cosmetic, and chemical industries. These industries have employed supercritical extraction in a variety of ways to improve the quality of their products and processes.

> *Did marriage or children impact your career?*

Patrick: I don't believe that my marriage has had any significant impact on
my career. However, my marriage has served as a haven in which I can retreat
and feel safe and loved regardless of career or work issues. My husband and I
have no children.

> *Has the marriage been very supportive?*

Patrick: I have a very supportive husband. I've always had a very interesting and delightful personal life. I've never been quite the stereotype of a female engineer or scientist who is "all alone." Loneliness has never been an
issue for me. I have always tried to separate my personal and professional
life. Whatever I do in public is one thing, when I come home I'm just little

ol' Jennie. I want my family and my friends to view me as a normal everyday person. I don't want constant attention in my private life or people prying into it. I'm not the kind of person who enjoys a lot of attention and gets a lot of satisfaction from it.

I like what you said because I do think that some have the notion that we are all lonely black women out here doing science. Tell me about your hobbies.

Patrick: Actually I can't imagine being lonely. It's not part of my makeup. I do spend a lot of time alone but I actually choose to be alone. I'm an avid gardener. I can spend 12 to 16 hours in my yard gardening doing hard labor. I really love nature and the outdoors. Most of that time which I spend alone is really quality time for me.

What is your husband's profession?

Patrick: My husband's area of expertise is internal medicine. I must say that I'm very proud of him. I am proud of him not just because of his academic achievement but more for his honesty, integrity, gentleness, and character. We are the very best of friends. Medicine is a second career for him. He had previously worked as a chemical engineer. Once I became ill he decided to go to medical school to be in a better position to support and understand my medical challenges.

What advice do you offer women who are seeking to balance marriage and job responsibilities?

Patrick: Be loving and supportive of the man in your life. Don't be afraid to reach out and touch or interact with someone who is not quite as successful as you and help bring him along. He might make a wonderful husband. I think we need to pull away from the traditional standards that are set by other people and look at our own. For instance, what are the particular circumstances that surround us as a black people? Black women are some of the strongest individuals on earth. We need to recognize that. Use your strength to better your own personal situation. Don't allow your strength to become a hindrance to your personal happiness. Be wise and thoughtful in how you deal with your private life. I often tell black women in particular (and people in general) that it is important to utilize one's brightness and intelligence wisely not only in your professional work, but also in your private life. A successful career is by no means the essence of life. It is important to keep a balance in one's life.

That's wonderful! Where would you like to be or hope to be in the next ten years?

Patrick: I hope that in the next ten years that I will have resolved my health challenges. Also, I would like to develop a business that will allow me to utilize all the knowledge that I have gained about environmental hazards and concerns to help protect the public. In addition, I would like to develop a mass media approach to reaching large numbers of African Americans about issues that affect our mental, physical, and emotional well-being. Hopefully within this time frame I will have completed a number of books.

What do you think is the most important advice you can offer an African American female hoping to enter science or engineering as a career?

Patrick: I think that's a very tough question, but I would first tell her to be herself. You need to know who you are, you need to be comfortable with yourself, you need to love yourself, and you need to respect yourself. Then everything else becomes secondary. Achieve the highest goals possible but don't allow achievement alone to define who you are. Make that decision early. Don't let material things or world recognition be your driving force.

Interview Dates: July 1996 and August 2003.

Selected Publications and Research Activities

Patrick, J. R., and R. C. Reid. 1981. Superheat-limit temperature of polar liquids, Ind. Eng. Chem. Fundamentals.

Patrick, J. R. 1981. Supercritical extraction technology. National Organization of Black Chemists and Chemical Engineering Proceedings.

Patrick, J. R., and F. Palmer. 1985. Supercritical extraction (SCE) of dixylenol sulfone (DXS), Supercritical Fluid Technology (Elsevier).

D'Souza, R., J. R. Patrick, and A. S. Teja. 1988. High pressure: Phase equilibria in the carbon dioxide-n-hexadecane and carbon dioxide-water systems. Can. Journal of Chemical Engineering.

Patent Disclosures

Supercritical Extraction of Chlorinated Biphenyls (PCB's) from Transformer Oil (10-C), Nov. 8, 1981, RD-13, 199.

Supercritical Extraction of Dichlorodimethysilane from Trichloromethylsilane, Jan. 16, 1981, RD-13, 377.

Method of Improving the Quality of Lexan-Polycarbonate Resin Made from Interfacial Polymerization, Nov. 23, 1981, RD-14, 050.

Purification of Bisphenol A (BPA) via a Supercritical Fluid Mixture, Dec. 7, 1981, RD-14, 112.

Use of Acetone for the Purification of Lexan-Polycarbonate Resin, RD-14, 231.

JANN PATRICE PRIMUS

Her Voice Lives On

WHEN I ARRIVED ON THE SPELMAN COLLEGE CAMPUS in the fall of 1997, I was met with a unique situation: I had my choice of black women scientists to interview. As I walked across the pristine campus in a sea of young, beautiful black faces, I saw a stately, confident, graceful young woman over six feet tall standing near the science building. As I moved closer, a wide smile appeared on her face and she extended her hand to welcome me. There was no doubt in my mind that Jann Primus and I would have a stimulating conversation on black women scientists. Jann did not disappoint me. We had a soul-stirring conversation that I will treasure for the rest of my life.

From Jann I learned that she had come from a long line of "movers and

The late Jann Patrice Primus, an associate professor of biology and the interim director of the Research in Minority Institution Program at Spelman College, was an outstanding young scientist and daughter of America.

shakers." It was not until after her untimely death in 2002 that I learned more of her family's rich and vibrant legacy in America history. With the urging and the support of her former colleague and friend, Shelia McClure, I contacted Jann's family about proceeding with the publication of Jann's interview. On a hot August day in Alabama a few months after Jann's journey to the other side, I received a historical treasure trove from Jeanne Johnson, Jann's older sister. This material included a copy of the 1711 land deed on the property held by Jann's great-great-great-grandfather.

Around 1711 the first Primus (then known as Primas), named John, arrived in the "new world" as a free slave and owned land in Charleston, South Carolina. John Primas's ancestors had sailed to the United States from Trinidad. According to family historian and political scientist William Primus, III, Derrick Primus, the great-great-grandfather of Jann and Jeanne, was born in 1880 and was quite prosperous. Although some links and details of the family history are missing, we know that the Primus family had several generations of successful citizens throughout America. Jay Hugh Primus (their father) was one of the surviving Tuskegee airmen and one of the first black dentists to practice in northeastern Ohio. However, this is just one side of Jann's family legacy. Jann's mother, Ruth Boston Primus, was the daughter of Alexander V. Boston, a graduate of Meharry Dental School, who had settled in the Appalachian town of Virgie, Kentucky, where he dedicated his life to serving the health and well-being of all people. When you take into account the great legacy of both the Primus and Boston clans, it is no surprise to see how we get a Jann Primus.

Born on May 17, 1959, to Ruth Boston Primus and the late Jay "Doc" Primus in Elyria, Ohio, Jann was the youngest child of three. She had an older brother, Jay Jr., and an older sister, Jeanne. Early on, it was clear that "baby Janny" was a special child. She always liked to do well at whatever she attempted to do. Jeanne fondly recalls Jann becoming very distressed when trying to learn to spell her middle name, Patrice. Big sister Jeanne told her, "Just put the words, 'Pat' and 'rice' together and you will have your name." At the age of six, she got her first library card from the Elyria Public Library. She read voraciously, completing about twelve books a week. Janny began her spiritual journey early when she joined the Asbury United Methodist Church in Elyria. This association continued throughout her adult life in the church affiliations she maintained wherever she lived.

Jann graduated from Elyria High School in 1977 and entered Spelman College, the same college her older sister Jeanne had graduated from two years earlier. Always outstanding in academics, Jann graduated summa cum laude from Spelman in 1981, the Centennial Class, with a major in biology. Jann continued her education at the Massachusetts Institute of Technology, obtaining a Ph.D. in biochemistry in 1987. At MIT, Jann excelled in her graduate studies and worked under the tutelage of Gene M. Brown.

Jann knew that she had the spiritual gift to teach and returned to her alma mater, Spelman College, to begin her career. She progressed through the ranks to become an associate professor in the biology department. From 1989 to 1992, Jann was a postdoctoral fellow in genetics and molecular biology at Emory University. She also worked in the laboratory of Victoria Finnerty on a Ford Foundation Postdoctoral Fellowship and a National Research Service Award from the National Institutes of Health; this collaboration spanned more than ten years. In May 1996, she Jann awarded tenure and promoted to an associate professor.

Recognized nationally as an innovator of science pedagogy, Jann served as a workshop leader/presenter at the Council on Undergraduate Research Institute at the University of San Diego, California in 1998. She was named by Project Kaleidoscope, a national organization plotting the course of undergraduate science education, as a member of the "Faculty for the 21st Century". In 2001, she was inducted into the most prestigious scholarly honor society, Phi Beta Kappa.

Jann was a member of several national scientific and civic organizations, including the American Society for Biochemistry and Molecular Biology and the National Organization for the Professional Development of Black Chemists and Chemical Engineers. Beyond her active professional life, she was active in the Cascade United Methodist Church and served as a volunteer for the Intervarsity Christian Fellowship Ministries—Southeast Region.

Her friend Lynda Jordan fondly remembers their graduate days at MIT when they would sit up all night doing homework and talking about their futures in science. Lynda remembers that the two things Jann loved were God and science. Lynda says, "she was an excellent scientist; however, her love of science did not compromise her love and relationship with God." Jann and Lynda remained great friends throughout her professional life.

Jann touched lives in so many ways, including my own. Shortly after I received Jeanne's letter I traveled to Atlanta to meet her. We met on Spelman's campus to talk. On that day Jeanne had a difficult job: she had to collect her baby sister's belongings from her office. I was happy that I could offer a little support for her, as Jann had done for many others. In her honor, Spelman has named one of its laboratories the Jann Primus Molecular Biology Laboratory, and a scholarship has been established in her name. Beyond being an outstanding scientist and educator, Jann was an exceptionally kind and loving human being. I hope her story and family history will inspire the next generation of scientists and engineers to always aim to be their best. May her voice live on through this interview and her legacy to education live on through her students.

How did you first become interested in science?

Primus: I think, in general, I had very good teachers. I grew up in Ohio, and I was interested in all the various areas until I was in the tenth and eleventh grades and I took advanced placement courses in biology. I had very excellent teachers, particularly in biology. I had a lot of opportunities to do extra lab work, etc. I think that was when I thought I might be interested in a science-based career.

Was there any particular person who influenced or encouraged you?

Primus: Yes, my advanced placement biology teacher, William Pearson, who was just a very good teacher. He would take us out to do fieldwork. He also had a part-time job working in a hospital laboratory, so he taught us things like how to do blood counts, how to test hemoglobin, things that were really more advanced than what we do with college students now. I thought that was just really exciting.

What role did your family play in your selecting science as a career?

Primus: My family just basically encouraged me in my academics. I think anything that I would have chosen to do they would have been supportive of me. I found my parents just generally supportive.

Did geographical location have anything to do with your interest in science?

Primus: I think that I was very fortunate. I lived in a very integrated community. The schools were very good for a mid-sized town of about 60,000 people, Elyria, Ohio. It's a suburb of Cleveland, and I think that the high quality of the academic program definitely made a difference. My high school chemistry and biology classes were as good as my college classes. I had very good teachers across the board—foreign language, math, and science classes.

I think the teachers really pushed me. Likewise the library and the music program were very good where I grew up, and the fact that I was in a very integrated community. Usually when you go into integrated communities, you have very good public schools. I think that made the difference.

> *Were your parents native Ohioans, or were they traditional in that they migrated to the North?*

Primus: It's a very traditional town and it's a traditional story of African Americans migrating to the North. For example, this town is ten miles away from the town where Toni Morrison was born, Lorain, Ohio. And the people who live in these towns were very traditional. African Americans came from Alabama and Georgia to work in these steel mill towns. My parents came from Georgia and Florida. It turns out we had family members who had gone there for industrial jobs, although my father's a dentist and my mother's a social worker. Other members of my family came from the South and settled in that area. My father established a dental practice in that area because there were a lot of African Americans who wanted a black dentist.

> *Do you feel race and gender played a particular role in how you are viewed as a scientist? If so, can you cite examples?*

Primus: I absolutely say both race and gender played very strong roles in how I've been viewed as a scientist. If you look at my name on paper, you can't really figure out whether I'm black, white, or whether I'm Swedish. In some respects, you can't even tell whether I'm male or female. I've had several experiences where I had an appointment that's scheduled at a university or company and the scientist that greets me opens the door and is surprised to see this tall black woman there. I'm thirty-eight years old, and if you've been black for thirty-eight years and a black woman for thirty-eight years, you know that look when someone is surprised and tries to recover. I've had that experience many times.

I think there are certain expectations for what my background should be. I have had the experience, particularly with older scientists—older white scientists, usually male—that they expect that I've come from a very impoverished inner city background. When I describe that I went to Europe when I was in high school or that my parents were professionals, they're very shocked. And in some cases, they seem to be angry that I am not the stereotypical black person that they would like for me to be. This is the first time I've ever mentioned that. I've had that experience with regards to being a black woman scientist and facing the fact they're surprised to see me. I often find those conditions even with very young white scientists, particularly those who haven't had experiences with African Americans, as many scientists don't. Since there aren't that many of us doing science, I've had the experience where individuals have talked down to me even though they've been told that I'm a molecular biologist. I personally find it amusing. It's just a game. To me what I do is I sit there while they're explaining their project in a kind of patronizing way telling me what this and that is. They know I have a Ph.D. from MIT. I am thinking, "I know what you are talking about." I find that if I ask a very high-level question or just communicate my understanding of certain concepts and my views on the subject, I get a totally different response. All of a sudden they realize, "Oh, she understands what I understand." After that, it's usually OK.

> So you feel like once again you have proven that you are worthy to be in the science academy.

Primus: Precisely. You see a look of surprise in their faces. In some cases, it's like they're very happy. I've had this experience like, "Oh, gosh. I never thought that this black woman could really know about cloning a gene," and it's OK after that. I figure the first time it happened to me that it may have upset me. Now I actually think it's funny because it can be used to your advantage too. You can open up dialogue and collaboration.

> Is it difficult to distinguish between race and gender issues?

Primus: I don't spend a lot of time thinking about the distinction between the two. My personal off-the-cuff view is that for me race has made more difference than gender. I have white female friends who are scientists, and I know that there are issues that they have too with regards to respect. But I think that there is more of an expectation of a black person not knowing what she's doing necessarily than a white female person.

> Are there factors other than race or gender? You've already touched on it when you were talking about your family background. One that I would like to bring up is your height, because you are tall and young. Has that made a difference?

Primus: I have had experiences on the positive side—and this is something I have mixed feelings about. For a black person in this country I come from a relatively privileged background. I really do. I've had mixed feelings about that in some of the white environments I've been in. For example, as a postdoc at Emory, one of the professors walked into my lab and he heard me listening to classical music. From that point on that particular faculty person appeared to really like me. I think I also saw that same person at a symphony concert downtown. I think for white people to see a black person who listens to classical music and talks about her favorite composers, that somehow made me an acceptable person. I think that my background meant that I was comfortable in certain cultures because I grew up in a predominantly white environment. I can speak the "lingo," and that's been good for me.

But on the other hand, I think sometimes it makes me a little uncomfortable because I think about those who do not have the same background as me and I wonder if they would be accepted the way I am. I think this class issue is something that comes up a lot, and for me it works in my favor, but I think for a lot of African Americans it works against them. And quite frankly, it doesn't hurt me to be a tall, confident black woman. I can look directly into the eyes of my colleagues.

That you can definitely do, Jann. So what brought you back to Spelman?

Primus: It's a family thing. Every member of my immediate family is a college graduate, and every member of my immediate family is a graduate of an HBCU (Historically Black College or University). I have a sister who's six years older than I, and she's a Spelman graduate, and we have maybe ten people in our family—cousins, aunts—who have attended Spelman. I went to a very large, predominantly white high school. When I went into college prep courses in my high school, there were very few blacks in the college prep courses, frankly. My school environment probably from the ninth grade on was predominantly white, even though it was an integrated school. I think for me socially and otherwise, it was really a good thing to come to this kind of environment, and although my parents would have supported me wherever I went to school, I am a proud Spelman graduate. My sister went to Spelman, and it was a very special place for me to visit. It was a place I really wanted to go to after high school.

You received your B.S. in biology from Spelman?

Primus: Right, and I have my Ph.D. from MIT.

So you went straight through from bachelor's to Ph.D.?

Primus: Right.

Do you feel the civil rights movement played a role in your success or in your selection of science as a career?

Primus: Wow! That's a tough one. I don't think so. I have a grandfather who's a dentist, and my father's a dentist. I have an aunt who's a nurse with a Ph.D. I really had the advantages of about everything. Even the high school I attended has been integrated for seventy years. I'm sure there's a connection, but I think my particular cultural background sort of shielded me from a lot of things that many African American women might have had to struggle with.

What about the women's movement?

Primus: I think so. I think the women's movement may have impacted my choices more, because I was free to choose a science career. I think about my family. My sister's a very bright woman who graduated from Spelman, and she's a schoolteacher here in the Atlanta area with three kids and a husband. My mother is a social worker and a graduate of Tennessee State University. I think I'm kind of a next generation in terms of someone who did not immediately think when I was twenty years old that I had to get married and start a family right away. So I think the women's movement probably had a bigger role in my family feeling comfortable with me and the choices that I made early on. For example, after I finished at Spelman I went to Boston at age twenty-one to live on my own.

How do you feel other scientists can be more supportive of black women scientists? What responsibility does the black woman scientist have in her own success as a scientist?

Primus: That's a really good question. I had excellent preparation at Spelman. I went to an environment that was different from Spelman. At MIT, clearly being in a predominantly white and male environment, I had some interesting experiences. In the graduate lab, I was the only female graduate student, and there were about four or five white male graduate students. I found out that there had been a black female graduate student in my lab several years before who did not complete the program. This was several months after I had been in the lab, and in a passing conversation someone said, "Oh yes, we had a black girl in this lab before." No one had said anything to me about it. And I think I felt that there was something hanging over my head. "Well, she did not make it in the program." So I wondered if they expected that of me. I even also felt that sometimes there were comparisons they made because again they would talk about this particular person playing loud music in the laboratory, and there was some sort of expectation for me to do the same. "Well, you're different than she is." I never met this woman, but somehow I felt the same thing I mentioned before, like "Well, you're an acceptable black person." I wished someone could have said, "Hey, listen. That person had her own issues. It had nothing to do with her being black per se. Let's not judge or use one size to fit all."

How many blacks were in your department?

Primus: At the time when I went to MIT, I was the only black in my department out of 110 to 130 graduate students. I was the only black—period. And the next year, I think two more came in that were successful and did finish.

When did you attend MIT?

Primus: I began in 1981. I was motivated and I knew that I either wanted to go to graduate school at Harvard or MIT. It didn't make any difference. I never thought about saying, "Do you have any black students?" I knew there were black students there because I had talked to recruiters. But I think if someone had said, "Hey, there may be some issues because you're different or might be perceived different because you don't go to beer parties or play loud music," I think that would have made it easier for me to deal with some of the issues that I encountered.

I've talked to several MIT graduates: Jennie Patrick, Shirley Jackson, and Lynda Jordan. Each of you seems to have had your own particular experiences, but it is interesting to me that MIT contributed to producing some outstanding African American women scientists.

Primus: Lynda and I were very much a support for each other. I am also a biochemist, so we took some classes together. We stayed up together and studied hard. I think if I come off as a strong black woman, Lynda really comes off as a super-strong black woman. I am a wimp compared to her, but we each have our own particular style. There is something about us—that we don't, as they say, "take no stuff." Both as a graduate student and as a postdoc I think my strength and self-confidence have definitely helped me more than hurt me. I was not afraid to confront issues and nip them in the bud.

I think actually being a student at Spelman helped me with that. Spelman built enough confidence in me to handle things that I saw happening to some other black graduate students where they would doubt themselves so much they would let things get to them. I think I had enough self-confidence that I would confront them, and I knew that I was just as smart as some of these other people. I actually found out that some of these white guys were just nice people who lacked exposure. They had the same concerns that we all have as graduate students. Some of them did not pass their general examinations the first time. Well, I did, but I didn't think that because they didn't pass they were somehow less than anybody else. There are a number of reasons why these things happen. So I think my confidence was intact long before I arrived at MIT. It helped me, and I was also sensitive to other students who had some of the same concerns that I had.

I never really answered your question about what we as black women can do. I have to say personally—and this is something I try to give to my students who I know will be going on to graduate school or medical school— I think it's really important to not project onto people your inadequacies or lack of confidence. I've talked about some difficult circumstances I might have gone through matriculating through to the Ph.D. and postdoc but I would say that the vast majority of experiences I've had with mentors and colleagues have been very positive. My experience has been that if you know your stuff, people really don't care. Once they know you've got something to offer, then the rest is history. You should really enjoy doing your work, number one, and be prepared to work very hard.

> So you are saying, "Don't project your lack of confidence on others when you are assessing whether or not you are experiencing a problem related to your race or gender"?

Primus: Exactly. And I think that is what happens when you have a low sense of personal confidence. You start seeing people minimalizing you when it may not be there. The most important person in the world, in terms of your destination, is yourself. And if you know you've worked hard and you know you've done what you've done, then you can't start imagining things about what other people think.

In my whole academic career, I don't think I've ever thought, "Does this person think I'm smart?" It just never crossed my mind. It can really make you miserable to worry about how other people in your department might feel or think about you. I mainly compete with myself. Another thing I found that it is not helpful as a graduate student or faculty member is to get into competitive things with other students or colleagues. You find your own project. You find what turns you on and you do that. In the end, you will be highly ranked, but you won't be so twisted up and tied up in negativity. Generally, what I have done is find my own niche and create opportunities to bring out my best as a researcher and instructor.

What do you say to people who have negative experiences that are not their imagination but may, in fact, be based on their race or gender?

Primus: I think that's a very good point. I have friends who have had those problems and moved on from them. If you have truly had negative experiences and it has come from those around you, then I would still say that having a good feeling about who you are and surrounding yourself with people who will support you are the best deterrents. You can call it the way it is, which is that someone is doing something to keep you from progressing and you have to realize that that is the other person's problem. You try to be truthful to yourself. I think it's really important when those negative things happen that you have enough confidence to move on if the situation is too difficult. I frankly have seen people really destroyed. I have a very good friend who is too concerned about what other people think of her. I mean, who cares? I'm more concerned about whether I respect what I've done myself. How does my family feel? How does my community feel? You have to put other people's opinion of you in the right perspective.

How has being single affected your life as a scientist?

Primus: I would say there are really two sides to it: the good side and the bad side, or the clouds and the silver lining. I think that I have been much more productive because I don't have a husband and children. And there's no way, thinking back, that I would have been as productive. Even now, though I don't work the same hours I did as a postdoc, I probably get more work done as a single person. So I think the fact that I don't have a husband or children has been helpful, in terms of the focus, productivity, and projects, collaborations, etc.

On the other side, I think that it is sometimes helpful to have an easy distraction from your work. I have a goddaughter. I have nieces and nephews, and they help me to forget about work, and I love to do things with them. I think when you have children or a husband or another life in that respect, it's a little easier to do that. Now, as a single person, of course, you work on your social life, but I think black women, white women, black men, and white men who are single today will tell you that sometimes the single social life is as much work as the professional life. I guess for single people the grass is greener on the other side. You don't have to construct family or social life, supposedly, so that ready diversion is there, but you have to balance them. Sometimes balancing a family with a life in science can be difficult, especially for women. So there are two sides to both situations.

That's true. I can relate to what you have said as a single person as well as the issues that my married friends deal with. But basically you have extended family here in Atlanta, and that makes a big difference.

Primus: That makes a big difference. That was a large portion of the reason why I chose to settle in Atlanta. I have my parents and a brother in Ohio, but the fact that I have classmates, friends, and also my sister and her family here in Atlanta was an important part of having that support system.

How long have you been here?

Primus: I came to Spelman right after my Ph.D. in 1987. But I left for three years, from 1989 to 1992, to do a postdoc at Emory.

Because of the long hours and rigors in science and engineering, young women often view being a scientist as having to choose between a personal life and a professional career. Is there any advice you would offer a young woman, especially a young black woman?

Primus: I don't think I'm in a position to give advice. I think you just live your life. I did not set up in my mind that I was making a choice to have a family or a career. I guess my views of family are different. I think nowadays black women and white women, if they have a certain level of professional commitment, are marrying much later. I really don't think it's a bad idea. I also know my personal emotional maturity. I really think that I was five to eight years behind where I probably should have been. If I had gotten married at twenty-three, I personally think it would have been a disaster. It wasn't until I was about thirty years old that I was even thinking maturely in that direction.

I ask this question because when I give these workshops and talk to young women, especially young black women, who hear all these things about the number of marriageable men or how career will affect their personal lives, they have this idea of choosing one or the other. Granted, more women are choosing careers and marrying later, but what do we tell our young women who are looking for role models?

Primus: I think if you talk to Spelman students, if you say, "I really want you to be committed to your career," they will say, "But I want a family." And I guess if they look at someone like me—I'm thirty-eight, I don't have children, I'm not engaged, I've had serious relationships—then they say, "If I choose that route that means no family." I don't see it that way, because there's a Shelia McClure who's done the same thing, but she has a husband

and one child. It might be hard to have five children and do what she does. She is a role model who has both. I guess my thing is that you really have to follow your heart. I like to live a more affirmative life, which means if there's something I really want to do, I do it. I don't allow fear to keep me from doing something. My personal view is, who knows? At age forty I could get married, at age fifty I could get married, and I can adopt a child. So today there is more than one role model for young women to follow.

> *Actually one of the women I interviewed married at forty-nine. You never know what may or may not happen or when it will happen in your life. That's why I want all views from black women scientists—so that our young women can see a range of experiences.*

Primus: Sure. I really do love children a lot, and I believe that if I do not have my own children I truly will adopt them. It's always been in my heart. I really would like to be married, but the right person hasn't come along. My view has been that as I achieve more and progress more, I meet better people. So if I had decided at twenty-one not to go to graduate school, then who knows, maybe I'd be in a low-level job someplace, not married, and not making any money! So I think it's better for me to make a positive choice to do the things that I want to do, and good things will happen, too.

> *I think that's a beautiful thing for young women to hear. What can we do as black women to change our invisibility and our image in science? How do we become visible?*

Primus: That's really a good question. I think that it's a very personal question, because it really hits where I am in my career right now. I recently was promoted and got tenure here. Now I'm in the third year of my first big grant, and it's going really well. I've been very blessed. The research is going well. It's a lot of hard work. I've got a couple of collaborations going on with people. A challenge I'm dealing with is being willing to take visible positions. For example, in national scientific organizations, we have to be willing to go for leadership positions, and that's where I am now. I think as we do that on an individual basis that we will have that visibility. I think as African American women we tend to have a certain view of service and we want to really serve our communities, but we are also scientists. I think it's really important not to just get relegated to the trenches and sort of get stuck in that mode of just doing science and teaching. They are both important factors in our lives as scientists, but we have to become more active in our own visibility. We have to have a more strategic view about where we're going.

Although I wouldn't have said so fifteen years ago, I think I am a very ambitious person, and I'm not ashamed to say that. It's a good thing to use the talents that have been given you. Recently, I was working with the search committee for the new president of Spelman, and I started realizing that we've got to keep moving, we've got to look for new opportunities. These are things that are involved in moving up the ladder, and this experience allowed me to observe other black women in administration and some of the strategies that they use to further their careers and lives. It was an eye-opener for me.

I haven't had that leadership role model for me. My mother was primarily a mother and wife, even though she did work outside the home. She was not a career person. Likewise, my sister is a schoolteacher, but she is not really a career person in the same sense that you and I have careers. She doesn't travel for her job, etc. My mom and sister have not had to deal with some of the issues that we are faced with. I think we don't necessarily have enough models in our age range to realize how to make those decisions and maneuver through the academy.

How do you envision your future as a scientist? Where do you see yourself in ten years?

Primus: I would imagine within the next few years I probably will be a department chair here at Spelman. Frankly, for any kind of advancement I make, I think I have to go through some administrative training. I doubt that I will be a bench scientist for the next twenty years. I have a very good friend at Emory who's a faculty member and is close to sixty years old. She's still wearing her tennis shoes at the bench. I don't see that for myself. I think maybe for the next ten years I will be doing research. I also see myself advancing in an administrative office. I can see two directions in my career. I can see becoming a dean or provost some place and being able to really direct and help some college or university develop a science program for teaching. Other possibilities include administrative roles in the National Science Foundation or the National Institutes of Health or somewhere in the biomedical establishment. I definitely do not see myself as just a bench scientist or just a faculty person; I think I can really effect change in other ways with my scientific background.

What do you think will be your greatest contribution to science?

Primus: I do a number of things. One of the things that I've done for the department here is coordinating our outreach efforts for several years. I do yearly workshops for high school teachers here in the Atlanta area, and that's a really positive thing. I really enjoy it. I have done a lot of faculty development work leading workshops for high school and college teachers. I would say, clearly, the most important contribution I've made in helping the young women and men that I've taught here is learning how to be critical thinkers. To me, that's really what it's all about.

Could you elaborate just a bit on your research?

Primus: I am using *Drosophila* as a model system for studying eukaryotic deficiencies in pterin cofactor metabolism. This research uses classical biochemistry and present-day molecular biology techniques to determine the proteins and genes that higher organisms need for the synthesis of two essential pteridine cofactors, tetrahydrobiopterin and molybdenum cofactor. I concentrate on characterizing the enzyme aldose reductase in *Drosophila* and characterizing the molybdenum cofactor gene in this model. The work may help scientists gain a better understanding of several human maladies, including atypical phenylketonuria, diabetic blindness, and mental retardation due to the molybdenum cofactor deficiency.

Do you collaborate on this research with other scientists here or in this area?

Primus: Yes. Some of this work continues my postdoctoral research with scientists at Emory University.

What advice would you offer young women, particularly young black women?

Primus: I would use the words of the scholar/writer/teacher Joseph Campbell. He has one particular term that he uses in his writings, which is "follow your bliss." I'm interested in psychology as well. For me, you have to know who you are as a person and pursue things that are in line with who you are as a person. Don't pursue things because society tells you to do it, because your parents tell you to do it, or whatever, or because you want to make a lot of money. Generally people who follow their bliss make a lot of money. Bill Gates is clearly following his bliss. I think that is very simple advice. I think it's very healthy advice for life. If you find yourself in a job or situation that you really can't stand, then you need to do something about getting out. Likewise, for your personal life, use the same advice. I tell young women, especially, to not really settle for things for appearance' sake. Put yourself in situations that are good for you.

Beautiful! Thank you.

Primus: I needed to hear that, too.

Interview Date: October 1997.

Selected Publications and Research Activities

Sepiapterin Reductase and the Biosynthesis of Tetrahydrobiopterin in Drosophila melanogaster. Doctoral dissertation, directed by Professor Gene M. Brown.

Switchenko, A. C., J. P. Primus, and G. M. Brown. 1984. Intermediates in the biosynthesis of tetrahydrobiopterin. Biochemical and Biophysical Research Communications, 120, 753–760.

Brown, G. M., A. C. Switchenko, and J. P. Primus. 1985. Enzymatic formation of tetrahydrobiopterin in Drosophila melanogaster. In Biochemical and Clinical Aspects of Pteridines (N. Wachter, H. Ch. Curtius, and W. Pfleiderer (eds.) Walter de Gruyter: Berlin-New York. pp. 119–131.

Primus, J. P., and G. M. Brown. 1994. Sepiapterin reductase and the biosynthesis of tetrahydrobiopterin in Drosophila melanogaster. Insect Biochemistry and Molecular Biology, 24, 907–918.

Beard, C., M. Benedict, J. P. Primus, V. Finnerty, and F. Collins. 1995. Eye pigments in wild-type and eye-color strains of the African malaria vector Anopholes gambia. Journal of Heredity, 86, 375–380.

K. Kamdar, J. P. Primus, M. Shelton, L. Major, A. Wittle, and V. Finnerty. 1997. Structure of the molybdenum cofactor genes in Drosophila. Biochemical Society Transactions 25(3): 778–783.

Research Funding

Faculty Fellowship, Center for the Scientific Application of Mathematics, DNA sequence analysis algorithms, Spelman College, February 1995.

National Institutes of Health, Minority Biomedical Research Support Program, Tetrahydrobiopterin biosynthesis in Drosophila melanogaster, $313,569, 1995.

Biomedical Research Improvement Program, MBRS Supplemental grant for instrumentation, $126,277, co-investigator.

Dolores Cooper Shockley

It's a Family Affair

Not only was Dolores Cooper Shockley the first black woman to earn a doctorate in pharmacology—in 1955 from Purdue University—but she was among the first African American students to receive a Ph.D. in any field from that university. Shockley, who grew up in Clarksdale, a small town in the Mississippi delta near the Arkansas border, was surrounded by successful family and friends who inspired her to become anything that she wanted to be. Several of her cousins were in the health professions. Initially, she wanted to be a pharmacist because Clarksdale did not have a local drugstore for African Americans, and she wanted to fill this void in her community. Fortunately for Dolores, her mother had the foresight to send her to a Presbyterian school where she could gain a good high school education in the sciences, especially in chemistry.

From high school, she attended and graduated from Xavier University

Dolores Cooper Shockley, the first black woman to earn a doctorate in pharmacology in the United States, inspires her children and students to become scientists.

in pharmacy. Because many of her textbooks were written by professors at Purdue University, it seemed like an ideal place to pursue graduate work. Although an outstanding student, Dolores faced many of the harsh realities of racial and social injustices during the 1950s. She was active in student and community organizations during her graduate studies at Purdue.

By the time she successfully completed the doctoral program at Purdue University, Dolores had been bitten by the "research bug." She received a Fulbright Fellowship to the Pharmacology Institute in Copenhagen, Denmark, where she studied with the best and brightest in medicine and pharmacology, including a Nobel Prize winner. It was a wonderful and exciting time in Dolores's life. She was a young, smart, and enthusiastic scientist with high hopes and dreams for her future. It was also the eve of the modern-day civil rights movement and racial issues were beginning to heat up in America. The 1954 *Brown v. Board of Education* case had just been decided. Just across the state line in Alabama, the controversial Autherine Lucy case was being battled. Shockley recalls,

> I remember loving being in Denmark. The Danish people seemed
> to value and respect me. It was just the beginning of a terrible
> time in America. I remember the local newspapers all wanted to
> interview me about the Lucy case.

In 1957, Shockley returned to the U.S. with a wealth of knowledge and enthusiasm for research and teaching. She returned to Meharry Medical College. During this time, Shockley met and married her husband, William Shockley. A microbiologist, he understood her desires and the struggles of everyday life as a scientist. Meharry offered a lot of promise. Founded in 1876 as the Central Tennessee College to educate African Americans in medicine, Meharry also graduated its first pharmacy class in 1890. Meharry is one of two black colleges that offer comprehensive programs in medicine, dentistry, and health sciences in the United States.

Shockley begin her tenure as an assistant professor in the pharmacology program and eventually progressed through the ranks to full professor as well as becoming chair of the Department of Pharmacology. At the time of this interview, she was the only African American woman to chair such a department in the United States. Although she has encountered racism and sexism in her career, Shockley was determined to maintain a career and family life. Having four young children early in her career, she committed most of her paycheck to pay for childcare. Her commitment to her children and to her science paid off. One of her daughters is a chemist and one is an engineer. She also has a son who is an orthopedic surgeon. Her concern for her family has also extended to her students. After many years of struggling through the ranks, Shockley made sure training of minority students in pharmacology was a priority. In addition to carrying a full teaching load and an active research program, she has tirelessly sought and fought to train African Americans in the field of pharmacology. Shockley came from a strong family who inspired her to become anything that she wanted to be; she continues this legacy both through her own children and through her students.

When did you first become interested in science?

Shockley: When I was a little girl, I was always interested in experimenting with different things in nature, like making ink out of berries. I loved various chemistry sets. Anything that was creative and related to chemistry was interesting to me. I was in the third grade when my parents bought me a chemistry set.

It sounds like your family had a profound influence on your early interest in science. Would you elaborate?

Shockley: Yes, my family was very influential. All of my family was interested in me pursuing medicine or dentistry. When I got to high school, my mother knew that I was going to pharmacy school, so she sent me to a boarding school. Clarksdale (my hometown) did not have a good chemistry class. I went to school about 150 miles away from Clarksdale for my junior and senior years of high school. I had a chance to study with an excellent chemistry teacher who really helped to prepare me. This was a Presbyterian school, and most of the teachers were missionaries, white, and from the northern states. The school had been started by missionaries who wanted to help black children in the South.

Your mother was visionary. Did she also have an influence on your other siblings?

Shockley: My mother wanted us to be what we wanted to be. I had two brothers and one sister. We would always say my older brother would be a physician, I would be a pharmacist, and my younger brother would be a dentist. It almost worked out that way. My older brother became a chemist and my younger brother became a dentist. I was going to open up a pharmacy and practice in my hometown. I did become a pharmacologist, but none of us ever returned home!

Obviously, you had some influence from teachers due to your mother's vision about your future. How did geographical location influence your selection of science as a career?

Shockley: Yes, that is true about my teachers. I grew up in Clarksdale, a small town in Mississippi. Back in those days, the schools and supplies were very limited for the blacks. We got the leftovers from the white school. Most of what I did was learned at home, in terms of working with my chemistry sets. Geographic location and the time I grew up in were important factors. It was a segregated society and I was in a small, rural southern town, but my family overrode those issues to a certain extent. There were a lot of educated people in fields like medicine and dentistry.

Would you share some of your college experiences?

Shockley: I received a scholarship to go to Oberlin College, but there was no pharmacy school there. My mother didn't want me to go there, anyway, because there was no dormitory space. It was in 1947, right after World War II, and my mom had gone to school in New Orleans. So I chose to go to Xavier College in New Orleans, which had a school of pharmacy. Xavier was a Catholic college. Although I had never been around nuns before, I adjusted to that type of college experience and it was a great for me. I did very well in the program.

Were you the first black female in the school of pharmacy?

Shockley: No, there were quite a few in my class. In fact, we were one of the largest classes in pharmacy. Just to give you a bit of history about some of my classmates, Earnest Williams became the mayor of New Orleans. Rigel Gumbel, father of Bryant Gumbel of the CBS morning show, was our class president.

That is really interesting. How about your graduate program?

Shockley: I went to Purdue directly from Xavier. I received my Ph.D. in pharmacology. I was the first African American to graduate from Purdue with a Ph.D.; that was in 1955. I selected Purdue for graduate school because most of my textbooks were written by professors who taught there. I was accepted at eight graduate schools. I didn't know much about research when I went there.

What led or prepared you for your career in research?

Shockley: I had a professor at Purdue who was from Norway and he was telling me about an outstanding pharmacologist in Denmark, Professor Moller, whom I should consider working with after graduate school. I was interested in going to Scandinavia. I knew if I went there, I would probably have a good experience. I had written to Professor Moller and he agreed. So I sent an application and proposal. I went to Copenhagen during the summer. In fact, it was in July and I took a huge ship called the *Rotterdam*, and it took about two months to sail there. I stayed there for two years, from 1955 to 1957. It was really a productive time for me. I completed four publications. I worked with some of the best. Incidentally, I worked with the 1997 Nobel Prize winner in Medicine. We were co-authors on two papers together. At that time, he was an M.D. doing his Ph.D. at the Pharmacology Institute.

> *What were some of your other experiences as one of the few black people in Copenhagen during the 1950s?*

Shockley: Well, everyone was very nice to me. The Danish people were very friendly. At that time, they did not see many blacks in Scandinavia. During that time, blacks tended to go to France and Germany. I knew it was unusual because I would have these little Scandinavian kids walk up to me and touch me to see if my color would rub off!

> *This experience seems like a far cry from what was happening in America at that time. How did you feel as a person of color in a country where you were valued (at least as a novelty) compared to a country where the color of your skin could have meant your life was worth little or nothing?*

Shockley: It was the first time I ever felt free in my life. In 1955, things were still so segregated. I was in Denmark when the Autherine Lucy case came up and an end to segregation in the schools was declared. I was interviewed by all the local newspapers in Copenhagen about this case that was happening in Alabama. As a native Mississippian, I had seen my own share of segregation and racism. At least in Mississippi, whites put signs for coloreds and whites so you knew your place. In Indiana, there were no signs, but people had a way of expecting you to know the difference and to know your place. For example, when I was with my white friends from Purdue they would go ahead and seat us at some of the restaurants, but when I went with my black friends they wouldn't even seat us.

> *That leads directly into the next set of questions. Do you feel race and gender played a role in how you are viewed as a scientist?*

Shockley: Yes, it has played a role—but not just as a scientist. When I first arrived at Purdue, I stayed in the dorm. The university had a wing of one of the big dormitories for the women graduate students. At that time, there were five men for every woman at Purdue, especially in science and engineering. I had

been there for two years before more women begin to come in as undergraduates. That influx of undergraduates led to the graduate women moving out of
the dorm. I had met a young white woman who was studying to be a clinical
psychologist. We became friends and decided to be roommates. We wanted to
live in West Lafayette, where the campus was. No blacks really lived in West
Lafayette, though. We tried to find housing there, but everywhere we would
go they would all of a sudden not have any vacancies—even though they had
advertised that there were vacancies. Of course, it began to be very clear that
nobody wanted to rent to us because I was one half of the renters. I feel this
was due to my race and the general problems of the society at that time. We
had tried for several weeks without much progress. Finally, a physics professor
at Purdue agreed to rent to us. The location was ideal, because it was near the
campus, and that had been our goal. We even lived next door to the president
of one of the local banks. Interestingly, he would often invite my roommate and
me for dinner on Sundays. I think he wanted to show me how liberal he was. It
was an interesting time. So I definitely encountered race issues, but I don't think
it affected me in terms of doing my science or establishing a scientific career.

> Do you feel gender played a role in how you were viewed as a scien-
> tist?

Shockley: Yes, I think early on being a black woman was helpful to me. I
think it was very positive in my particular case. For example, I can remember
having two white female colleagues who wanted a Ph.D. in pharmacology
but they were literally stopped at the master's level. They were totally discouraged from pursuing the doctorate. But I was treated like I was just one
of the bunch. They felt that my white women colleagues were going to get
married and that would be the end of it. I guess in my case they didn't care. I
really don't know for sure. That was early on in my graduate career.

However, things really changed as I began my professional career. At
least two times, I believe that gender became an important issue in my career.
When I was offered the job at Meharry, the salary that I was quoted was ri-
diculous. I didn't hesitate to tell the president that I could not possibly work for
that salary. In fact, I told him that I would not work for less than the men who
would be starting out. My three male colleagues were offered the very salary
that I had quoted him. Because I was a woman, he felt that I did not deserve the
same salary as the men.

This type of behavior continued at different points in my career. My
former acting chair did a similar thing to me in regards to a raise that I deserved. He recommended all the males and one white female colleague be

given raises. He said that since I was married, I didn't need a raise. He was a white male telling me this. Gender has definitely played a role in my career as a scientist. I guess to some extent maybe it was race also, since he did want his white female colleague to get a raise. It's hard to tell, sometimes. In my case, being married was, in some ways, used against me. For a man, marriage was seen as a positive characteristic but not in my case. At this point, I am the first black woman to chair a department of pharmacology in the United States. I became acting chair in 1988 and I was made permanent chair in 1994.

> *Are there any other factors that influence how you are viewed as a scientist?*

Shockley: No, there are no other factors that I can think of.

> *Do you think that the civil rights movement played a role in your selection of science as a career?*

Shockley: Yes. Civil rights, in general, were very important. That was why I had a desire to become a pharmacist and open my own drugstore in my hometown. Blacks were not given access and resources to truly be successful. I was one of the lucky ones that had a family of professional people who had some access and privilege. The incidents and problems of my hometown made me move in this direction for success. Of course, I had finished my education by the time the modern-day civil rights movement got into full swing. I began college right after World War II ended, and I was about finished with my Ph.D. by the time the Brown v. Board of Education decision was made.

> *Do you feel the women's movement played a role in your selection of science as a career?*

Shockley: No, not really. I was on my way, but I think the women's movement has helped a lot of women become more aware of the possibilities as well as the obstacles. The glass ceiling is a real issue, for instance. Women's groups have brought that to the forefront.

> *What responsibility do you think black women have in promoting science?*

Shockley: We have a responsibility to participate in outreach programs where young girls might be turned on to science. They need exposure to careers in science and some of the black women who have science careers. I've worked with outreach programs in my church and programs here at Meharry. One of our programs is associated with the Links, which is a national service organization. Being associated with a national organization is a great way to spread the word and have programs for our youth in the community. I go a step further: in my church, I tutor children in math and science.

We have a responsibility to encourage our students. I am amazed at how minority students are often discouraged from pursuing science as a career. In the past, I have visited sites for various programs for minority students, and I am concerned that the attitude of some the faculty and staff at the large universities is very unconducive to minority students. It is such a cold environment. Some of these major universities have to be willing to understand and appreciate the sometimes disadvantaged background our young minority students are coming from. These students need a lot of time and encouragement. It's amazing what some of them can do, but they need a little more nurturing, time, and exposure. I believe that all of the students that I train should be able to go much further than I did.

What role did having a family play in your career as a scientist?

Shockley: I married a scientist. My husband is a microbiologist. He understood what being a scientist meant. He knew the demands of the research and the whole publish or perish game. I think I didn't have my total concentration during those early years. I had four children and that affected my progress. I would have done more, but it was practically impossible. The good part was that the children were close in age. So once they were older and more responsible it helped a lot. I had a full-time housekeeper to stay with them during the week. It took practically my whole paycheck to pay for the housekeeper, but at least I stayed active in my career.

Are there any particular strategies that you used that you would recommend to young women, especially black young women, who are pursuing a career in science?

Shockley: I think it is important to stay current in your field. If you have children, you have to find daycare or have someone reliable to stay with them. In those days, daycare hardly existed. It has really boomed in the last forty years. It also depends on the number of children that you have. As a fellow scientist, my husband was understanding, but he was not as helpful as today's modern husband. It was the times in which we lived. Most of the responsibilities were on me.

I suspect that you might find some young women still having those same experiences today. Along the same lines, do you think that young women have to choose between a family and a science career?

Shockley: I don't think she has to choose, but she needs a mate that understands her situation. I am concerned that young women of today have a hard time finding young men at their same level. In our Department of Pharmacology, we have seventeen graduate students: four are male, and the rest are females. There aren't very many black males entering graduate school, and the numbers are also declining in medicine as well. Our medical class this year is 52.5 percent female. In the past, you had more males.

What can we do as black women to become more visible as scientists?

Shockley: I think we need to participate more in our national societies. I have been active in my professional societies. I was chair of the membership committee for the Pharmacology Society.

How do you envision yourself in ten years?

Shockley: I will be retired.

What do you think your greatest contribution will be to science?

Shockley: I would say being an educator. My greatest contribution would be educating health professionals and doctoral students. Before I became the chair, we were not really training graduate students. I have tried to educate and train graduate students. We were trying to graduate at least three Ph.D. students per year. We are well-funded as a department with grants because we have worked for these resources. The former chair was not very positive about our graduate training program. He was not encouraging of our great history and the opportunities for minority students. Although he was an excellent pharmacologist, our graduate education suffered during his tenure. As the new chair, I've tried to reinstate and strongly promote graduate education. About half of all the minority Ph.D.'s in pharmacology have come from our program. I think this will be my greatest contribution.

I think what you have done in the education of future pharmacologist is great. That's quite significant. Briefly, tell me about your research.

Shockley: I have been interested in the area of drug abuse. I have been looking at agents called pharmocotherapies for combating acute and chronic stimulants such as cocaine and amphetamines. One discovery we have made is a group of drugs called Ca^{++} channel blockers. These agents have potential for treating patients with cocaine dependency and addicts who overdose and are brought into the emergency room. We found that there is a certain level of benefit, but at a high dosage these agents become toxic. Other labs have confirmed our findings as well. Another area of interest is to learn more about how cocaine affects the brain's neurotransmitters.

That research sounds quite exciting and useful, because drug abuse is rampant in American society. In the future, I would like to know more about what happens with this research.

Shockley: Yes, the research is quite exciting. I will keep you informed.

What are your final thoughts and advice that you would offer young people about a career in science?

Shockley: They should never be afraid to try what they really want and never be discouraged. They should close their ears to discouragement and negativity from others. After they have done this, they need to study hard and be prepared. They have to take the time to get a good background in math and science. They must be literate and realize that they will be reading all their life. They will use math at some level in almost any science profession that they might pursue. So a good solid foundation and preparation along with a positive attitude are keys to a successful science career.

Interview Date: October 1997.

Selected Publications and Research Activities

Saunders, C. R., D. C. Shockley, and M. E. Knuckles 1997. Suppression of locomotor activity in rats after acute exposure to flouranthene. The Toxicologist, 36, 220.

Brown, V. L., T. Franklin, L. Williams, G. Woodall, and D. C. Shockley. Changes in stereotype, locomotion, and DAT binding after cocaine and cocaethylene administration. The Pharmacologist, 39, 88.

Twum-Ampofo, A., L. H.Wade, and D. C. Shockley. 1996. Changes in locomotor activity, core temperature, and heart rate in response to repeated cocaine administration. Physiology and Behavior, 60, 1–7.

Shockley, D. C., L. H. Wade, and M. M. Williams-Johnson. 1993 .Effects of alpha-2- adrenoceptor agonists on induced diuresis in rats. Life Sciences, 53, 251–259.

Ansah, T. A., L. H. Wade, and D. C. Shockley. 1993. Effects of calcium channel entry blockers on cocaine and amphetamine induced motor activities and toxicities. Life Sciences, 53, 1947–1956.

Crowell, B. G., R. Benson, D. C. Shockley, and C. G. Charlton. 1993. S-adenosyl-L-methionine decreases motor activity in the rat: similarity to Parkinson's disease-like symptoms. Behavioral and Neural Biology, 59,186–193.

RUBYE TORREY

Just Call Me a Scientist

A SOFT-SPOKEN PERSON, RUBYE TORREY grew up in Sweetwater, a small eastern Tennessee town, with a loving mother and grandfather. Because her father had died when she was just a baby, her grandfather served as important figure in her early life. She credits him with nurturing her love for nature and creating an enjoyable childhood for her and her sisters. As for her science career, teachers in high school and college played a vital role in encouraging her to pursue chemistry as a career.

A graduate of Tennessee State University, Torrey worked a while before pursuing a doctorate in chemistry. During this time, she also became a wife and mother. Her husband, who was also in the sciences, was pivotal in her decision to obtain a doctorate. Truly a pioneer in her discipline and profession, Torrey was the first African American female to graduate from Syracuse University in New York with a doctorate in chemistry. She worked at

Rubye Torrey, an assistant vice president for research and chemist at Tennessee Technological University, doesn't believe in labels.

various colleges and universities before accepting her current position as assistant vice president of research at Tennessee Technology University, making her the first African American woman to hold such a position at that university. Rubye, a woman of few but powerful words, has this to say about how black women are viewed as scientists: "We do not need to be viewed as black scientists, Negro scientists or African American scientists, we simply need to be accepted as scientists. All the adjectives are not needed."

How did you first become interested in science?

Torrey: I was naturally curious. I have always enjoyed watching the rain and wondering why and how it formed. I enjoyed watching the change of the seasons. So I have always been extremely observant of my surroundings and generally had the question why. What makes something work like it works?

Was there any particular person who influenced you or encouraged you in science?

Torrey: My mother strongly encouraged me. I had two teachers who were the driving force for me to go into the sciences, in particular in chemistry. One was a high school teacher. He was very impressive. He was extremely good in mathematics and chemistry and I enjoyed those classes with him. I decided then that I would like to pursue those subjects. In college, the chair of the chemistry department, Dr. Hill, was very influential in my choosing chemistry as a career.

Where did you grow up?

Torrey: I grew up in Sweetwater, Tennessee, about forty miles southwest of Knoxville.

Did geographical location have anything to do with your interest in science?

Torrey: I grew up in East Tennessee with access to a most beautiful environment. In trying to understand the beautiful world around me, East Tennessee was the greatest incentive that one could have to study science. We had a very large lawn and we had lots of trees and I enjoyed looking at them. My granddad was a farmer and I enjoyed watching the things he would do on the farm. He always took care of our garden. I enjoyed the horses, cattle, and chickens. I'm very fond of horseback riding.

Are you the first chemist in the family?

Torrey: Yes, I am.

Are there other siblings who majored in science?

Torrey: No. I kind of left the pack. I have two sisters, both of whom are older. I have one sister now because one has passed. My older sister is blind, but she is very interested in music. She is very good in music; she sings beautifully. My sister who passed was very good in French and reading. She was a counselor and second- and fourth-grade teacher, and taught French.

Did you have any brothers?

Torrey: I didn't have any brothers and my father passed when I was just a baby.

Where did you get your bachelor's, master's and Ph.D. degrees?

Torrey: I started my baccalaureate degree at a little boarding school, Swift Memorial Junior College. Swift Memorial was a little Presbyterian college. I am Presbyterian. I transferred from there to Tennessee State, where I earned a baccalaureate degree with highest honors. I remained there on a scholarship and got my master's of science degree. After getting my master's degree, I worked a few years before pursuing my doctoral degree. Later, I went to Syracuse University and earned a doctorate.

Once you got your master's, was getting your doctorate encouraged or was it always in your plan?

Torrey: It was always in my plan even before pursuing the master's, but I was certainly encouraged. I did very well in my coursework.

When did you get your Ph.D.?

Torrey: I got my Ph.D. in 1968. I was told that I was the first Negro female to get a Ph.D. in chemistry from Syracuse University in New York.

Has race or gender played a particular role in how you are viewed as a scientist?

Torrey: I am convinced that both race and gender have played significant roles in how I am viewed as a scientist, with race playing the greater role. Race and gender issues are very difficult to distinguish because there are so many similarities; however, they can be delineated.

Are there other factors, other than race and gender, that have influenced how you are viewed as a scientist?

Torrey: For more negative things, race and gender were tied into that one way or another. Because there's a stereotype about who can or should be a scientist, you can be affected directly or indirectly. When, as a Negro female scientist, you do not fit the stereotype, then you're just not someone that they can deal with. There may be situations where I am denied opportunities for other reasons; however, I would not know whether or not these were justifiable reasons.

Our acceptance as scientists is what is needed. We do not need to be viewed as black scientists or Negro scientists or African American scientists; we need to be simply accepted as scientists. None of the adjectives are needed. This is the link that is missing and must be supplied if we and the generations that follow are to reach the level of being all we can be.

> *I think that is a beautiful answer. You are basically saying we shouldn't label the black woman scientist. Please tell me more of your experiences with race and gender issues.*

Torrey: In the case of Negro females before all the big integration, you were challenged because people thought that you were in the sciences because it was fashionable to have Negroes in the sciences. So there are those who would like to challenge that because it was hard for them to believe. I even had people ask me, "Why were you trying to do that back then?" I think some people saw it as a fad because they were giving a lot of money at one time for minorities who were going into the sciences. And I said to them, "I didn't have to pay for my education because I got my scholarships based on merit and that wasn't a fad." If I were not a Negro, I am certain that I would have been perceived as a virtual storehouse of wisdom.

> *From that response, you seem to indicate that the civil rights and the women's movements did not play a role in your selection of and success in science.*

Torrey: My choice to become a scientist predates the civil rights movement and the women's movement; hence, they have not played a significant role in my decision to become a scientist. However, the atmosphere they have provided has been encouraging.

> *How have marriage and family influenced your life as a scientist?*

Torrey: Marriage played a significant role in my life as a scientist. I married and had two children after I completed my doctoral work. My husband is a biologist and biochemist. Early on, he was a medical technologist. He was helpful in many ways.

> *In what ways was he helpful?*

Torrey: We worked as a team. I feel this is the only way one can balance a family and career in science, particularly in chemistry. Set your goals, make your plan, and work your plan.

How do black women become more visible in science?

Torrey: We must do more mentoring as a group. We must work together as a group to be more visible and give strength to those who will follow us.

What is your area of specialization?

Torrey: My area is radiation and electroanalytical chemistry. My master's thesis was a project for the Tennessee Valley Authority. I developed a chemical method for quantitative determination of incipient spoilage of fruits and vegetables indigenous to the state of Tennessee. It was an aeration method that measured the quantity of volatile reducing substances in fruits and vegetables at different stages of maturity. It was determined that the critical value was 103 micro equivalents.

My doctoral research dealt with developing a mechanism for the alpha radiolysis of gaseous hydrogen sulfide, which was unknown at that time. I did some post-doctoral research at Brookhaven National Laboratory in the mass spectrometry division, where we looked at the pathways of the gaseous formation of certain noble gases using high-pressure mass spectrometry, which was just coming into being. Later, I was a visiting scientist at the National Institute of Standards and Technology (formerly the National Bureau of Standards), where I performed some of the ground research on acid rain.

Through the years, a real interest of mine has been the electroanalysis of drinking water for heavy metal content and the electroanalysis of human head hair for certain metals and their relationship to certain illnesses. I am still very interested in these areas and food chemistry in general.

What did you do before you went to Tennessee Tech?

Torrey: I taught at Tennessee State University for many years. I enjoyed teaching there very much. I started Research Day at Tennessee State University. The twenty-fifth anniversary was celebrated in the spring of 2003. It was started out of my desire to see the students experience the atmosphere of regional and national professional meetings. At that time, students were unable to attend such meetings. Only a few faculty were engaged in research and even fewer faculty were engaged in sponsored research. It has grown exponentially since that time.

How long have you been at Tennessee Tech? What was your motivation for pursuing a career in administration?

Torrey: I have been at Tennessee Technological University for almost sixteen years. I was hired as an administrator. I had been at the National Institute of Standards and Technology before coming to Tennessee Technological University and had been reviewing proposals for the federal government for many years. All of my experiences made me a natural fit for the administrative position I occupy. In the absence of a laboratory, my research interest is currently in the area of research ethics, which is a very hot topic now. I started and serve as chair of the Ethics in Science and Technology Division of the Tennessee Academy of Science.

What is the work in the Ethics Division of the Tennessee Academy of Science about?

Torrey: This section is very concerned with ethics being made an intricate part of research training and making it a part of the scientific curriculum.

Please tell me what a typical day is like for you as a research administrator.

Torrey: A typical day includes checking any new areas of research being announced on the web and distributing that information to appropriate faculty. I work with faculty who are writing proposals and help to process those proposals. I am involved in any duties that are related to proposal processing and grant writing. In addition, there are the usual meetings that I have planned and any impromptu meetings.

What do you think your greatest contribution to science will be?

Torrey: I feel my greatest contribution will be through teaching and mentoring.

Where do you see yourself in ten years?

Torrey: I hope I will always make a contribution. Perhaps it will be through writing and lecturing.

What advice would you offer young women interested in science careers?

Torrey: The greatest advice I could give is to make sure that being a scientist is what you want to do. You should make sure you are interested in making a contribution to science and society, and not just promoting self.

I have taught, guided, and counseled many young people over the years, many of whom are out there in the workplace. They were told, and I am still telling those who are coming along, that an investment in education will always pay great, great dividends. No one can take your knowledge away from you.

Interview Dates: November 1997 and September 2003.

GERALDINE W. TWITTY

Still on the Battlefield

AS I WAITED IN THE HALL OF THE ERNEST J. JUST Science Building, I knew
I was in the presence of greatness and history. I looked up and there appeared
around the corner a seemingly quiet woman who immediately apologized for
being a little late. As I was soon to learn, quietness is not a trait one would normally use to describe Geraldine Twitty. She made every moment of my wait
worthwhile. I listened intently as she began to educate me, share some history,
and passionately talk about her life at Howard University. I forgot a few of my
questions because she had so fully engaged me in her life story.

Born in Roanoke, Virginia, Geraldine was the youngest of six children.
Her father was a minister and her mother was a homemaker. She was raised in a

Geraldine W. Twitty, zoologist and biology professor at
Howard University, examines environmental justice for
all.

close-knit family and community where everyone nurtured and watched out for the well-being of all the neighborhood children. She recalls being challenged at an early age to do her best.

Geraldine began her freshman year at Howard University shortly after World War II ended. It was a jubilant time in the country. The war had ended and opportunities for African Americans and women were more open than they had ever been. Geraldine recalls that "it was assumed that females in science were at college to look for a husband and not there for the challenges of science and the career it would bring. Of course, I was there to do my science and I was frankly, footloose and fancy free." Unlike today, typically more young men would be found in science majors at the undergraduate level.

After receiving her bachelor's and master's degrees from Howard University, Geraldine accepted a position at Carver Research Laboratories at Tuskegee University. Her stay at Tuskegee was not very long. A chance meeting with a former professor, Margaret Collins, at a football game between Tuskegee and Florida A & M University (FAMU) led her to her next position. Geraldine recalls being completely surprised when Collins called from FAMU to offer her a teaching position. She accepted and worked there for several years. She also journeyed to UCLA for her doctorate. Though Geraldine worked hard to complete the doctorate, this did not happen. Although she had experienced racial discrimination in 1958, it did not stop her determination—it only delayed her goals. During this time, she also found the love of her life. Her husband worked with the Veterans Administration, which meant that his job required him to transfer to various locations for promotions.With two young children, they needed to settle their family and moved to Washington, D.C. It proved to be a great decision for the family as well as for Geraldine's career. She was able to complete her doctorate in zoology at Howard University.

Geraldine has been back at Howard for over three decades. She says she teases her freshman class by telling them about "when I was a freshman ninety years ago." Gerri (as she is often affectionately called) is unwavering in her commitment to her students. Having trained a range of students, she states, "We have to meet our students where we find them. We have to accept them with deficiencies but realize that they too, may one day become productive scientists or simply productive citizens." She has taught an array of courses and managed to conduct some research on tiny creatures called tardigrades, invertebrates which are notable for their capacity to carry out anhydrobiosis. In other words, these tiny creatures can lose significant amounts of water, survive and rehydrate themselves upon favorable conditions. Trained and mentored herself by the noted zoologist Margaret Strickland Collins, Geraldine has continued her work in zoology as well as branching into other territories. Her most recent work has led her to become an advocate for justice—that is, environmental justice—both at the academic and the grassroots levels. She feels it is tragic to not prepare more African Americans in the environmental sciences. She works diligently to bring the message to the Howard University community. Twitty is a person who will never give up; she will continue the battle to include more of the environmental sciences in the curriculum at Howard University and to push for the education of underprivileged groups in environmental causes in the larger community.

How did you become interested in science?

Twitty: There was a high school teacher who would not let me get away with being lazy. When I decided that the only way to avoid my mother knowing that I was being lazy was to complete the assignment, I found myself fascinated with what I was doing. I knew that zoology, and not piano, was for me.

Tell me how your teachers influenced you.

Twitty: I would say the better teachers definitely influenced me. As I reflect, I can think of a grammar school teacher who drilled us in English grammar and in reading and another grammar school teacher who introduced us to public speaking. I know that my comfort when speaking in the classroom or at a symposium comes from those early experiences.

For that reason, I have a real concern about what our children are experiencing in today's classrooms. The success we experienced can be attributed to a cadre of very dedicated black teachers who instilled in us the "you can do it" attitude. It is also important to recognize the strong influence of family and community. How naïve I was to complete high school and not realize that the next step for everyone was not necessarily college! Too few of our children today are exposed to teachers who demand excellence.

Were your teachers mostly male or female teachers?

Twitty: The high school teacher was a male and a good friend of one of my brothers. The elementary teachers were all females. In college, I was influenced by my general zoology teacher, who was a female.

How did your family or geographical location influence you in science?

Twitty: I grew up in Roanoke, Virginia. I was the last of six children. My father was a Baptist minister; my mother was a homemaker. As I look back on it, there was a tremendous amount of nurturing within the family. My dad died when I was four, so I only have mental pictures of him. My sister, brothers, and I were raised in a truly cohesive family setting. In a small town, everyone knows everyone else—my sister, my brothers, and I were expected to toe the line. There was a strong neighborhood influence. You were a child of the neighborhood. You were corrected by your neighbors, who were respected as members of an extended family. I don't think that sort of extended family is as well organized today.

Where did you go to college?

Twitty: I am a Howard University graduate at all levels. When I was an undergraduate I took my first course in zoology from Margaret Collins. I was offered an undergraduate assistantship under her tutelage. I was given a graduate assistantship in zoology as well. I left after my master's degree. I initially applied for a job in a hematology lab at the National Institutes of Health (NIH) based on my master's research in parasitology, which involved blood work, but I was offered a job washing dishes!

Later that same summer, on a visit to a brother in Tuskegee, I toured the Carver Research Foundation and was immediately hired. Several months later, I went to a football game between Tuskegee and Florida A & M University held at Tuskegee. At the game, I encountered my first zoology teacher. She had heard that I had received my master's degree and promised to remember me should she ever needed an instructor. It was a complete surprise when she called to offer me a job teaching the entry-level course at FAMU.

Did you take the job?

Twitty: Yes, I taught at FAMU for several years before I was challenged by a former physics teacher at Howard to start working on the doctorate. I applied for and was awarded an assistantship at UCLA. There I experienced what I now recognize as more sexism than racism. I retreated to FAMU, got married, and temporarily devoted myself to my family and teaching. We moved around a lot, so that made it difficult to further my education as well.

Was your husband in the military? How long were you in each place?

Twitty: We were three years everywhere. My husband was with the Veterans Administration. Promotion was tied to being transferred from various locations. We finally decided to accept a transfer to the VA Central Office in Washington with the idea that he could transfer to another agency rather than transfer to another town. With two young daughters, we felt strongly about the need for some stability in all our relationships, but particularly with respect to their education.

How long have you have been at Howard?

Twitty: I have been here since 1967. It seems that I have never left. I enjoy saying to my students, "Ninety years ago when I was a freshman here," and they respond, "I knew she was old!"

How did race and gender affect how you are viewed as a scientist? You alluded to some of your earlier experiences at NIH.

Twitty: Yes, my first real rejection was with the job at the NIH. Dishwashing 101 does not require a master's degree. That was racism.

At UCLA, my major professor never supervised my laboratory performance nor commented on the quality of any test I conducted. In the absence of any negative comments, together with the repeated re-appointments, I assumed that my performance was acceptable. My major problems came with what I perceived as unnecessary verbal attacks in seminar presentations and my inability to get him to approve what I considered a challenging proposal for research. It seemed that my comprehension of background papers in the field was never as extensive as those of the males who had been in the program for three or four years. My research proposals were always simply "too ambitious."

But how do you separate out the differences between what is race and what is gender? When did this happen?

Twitty: This was 1958 at UCLA. Not only were black professors in the sciences nonexistent, black students were also nonexistent. I was the only black graduate student enrolled in biology at the time. I considered changing major professors. The logical replacement that I was going to select was planning to leave at the end of the year. My assessment of the situation led to my abandoning the doctorate attempt and returning to Florida. This was not an easy decision, since I had completed my coursework, satisfied the language and oral examination requirements, and was ready to begin my research.

About a week prior to my departure, a white male graduate student stopped by the lab and announced that I was leaving for the wrong reason. He suggested that I probably thought my professor did not like me because I was a Negro, when the reality was that I wasn't liked because I was a female. It turned out that the professor lived in the shadow of the greater scientific achievements of his wife. He was known to have not approved a meaningful doctoral project for any female student. One after the other, females left the program. Thus I was the only female graduate student in parasitology. When one considers the double bind, I was certainly doubly bound. To a certain extent, the double bind will probably always exist. A black female always represent a dual minority status.

Was this the first time that you experienced the gender problem?

Twitty: No, there were gender problems during my undergraduate days when many of the male students were returning from World War II. I think that the attitudes stemmed from the idea that women in the sciences were looking for successful husbands rather than seeking a career for themselves. It was subtle and we tended to ignore it.

I did get my master's from Howard. In those days, the options for science majors were either medical or dental school or graduate school. I could not reconcile the demands of medical school with family life, so for me, graduate school was the only course of action.

How has your career been affected by race or gender? How do you think that you are viewed by colleagues and other professionals?

Twitty: In retrospect, gender bias has always occupied an underlying niche in the sciences; the sciences are male-dominated. From time to time there are issues that appear to evoke postures that are suggestive of racial and gender bias, but it is seldom possible to cite overt tendencies. It is sad, but the biases are often instigated by white females. One needs only to cite data from NSF or the Chronicle of Higher Education to substantiate such cases.

What is the number of faculty in the biology department, and how many are women?

Twitty: There are twenty-four faculty and four are women.

Describe the racial composition of the four women.

Twitty: Three of us are black and one is white. All of us have doctorates; most recently, our white female colleague was promoted to associate professor.

Are you a full professor?

Twitty: No, but I am tenured. That represents another interesting scenario. As a new faculty member, I was consistently assigned the largest classes. After several contract renewals, I was given tenure, along with a number of other females who had been similarly retained in the College of Liberal Arts. We should all be thankful for the Vice President for Academic Affairs, who was black and female, because she recognized what was happening to us. Subsequently, promotion requirements in the department were instituted that effectively made promotion unattainable for those of us who taught large classes and had families. Another factor involved in my situation was my attitude. I considered my contribution to the family finances to be supplemental to that of my husband, an attitude that lessened the drive necessary to sustain the years of struggle to full professorship.

So those promotion requirements really left some of you out of the competition for advancement. That sounds like an administrative issue. Were race and gender factors there, or was it just a general problem for new faculty?

Twitty: It stems primarily from race. I feel that a significant number of non-African Americans feel that we are fundamentally deficient.

Yes, but that's no longer an issue for a tenured professor, or do you not care at this point?

Twitty: Certainly, I care! We cannot allow rank to replace our concern for each other. We are where we are because those before us cared. Students are quite perceptive. One's demeanor outside the classroom is as important as that within the classroom. Student-faculty relationships must never suffer because a faculty member forgets why we are here. Without students, there is no need for faculty. Students must come to understand that our success rests on the extent to which we can provide them with skills and fundamentals to make them truly competitive. Rank should never impede or compromise our effectiveness.

What role do organizations have in your life as a scientist? Do you participate in any scientific societies?

Twitty: Absolutely. My participation is based on the need for continual growth, not to socialize or fraternize. The information flow in the biological sciences is such that it is almost impossible to stay abreast of what is happening unless you attend national meetings as well as read journals.

Did you have a mentor, and do you think mentors and role models, especially for black girls, are important?

Twitty: Mentoring was not something I was consciously aware of in my youth, although that is exactly what happened. In retrospect, Dr. Collins was indeed a mentor. She had a decided, positive influence on the career choices I made, without my noticing it. Even after the doctorate, we found common interests in participating in various scientific endeavors. She died suddenly in 1996 during a scientific expedition to Guyana. I was asked by her family to make comments at her memorial service. At that time, I came to fully appreciate the impact she had on my life. Indeed, making those comments at the service called forth aspects of our relationship that all spelled role model and mentor. My awareness of the positive influence she had on my life necessitates that I return the favor as often as possible.

As I reflect upon my interactions with students, I hope that I have been a mentor. I do have continued contact with a number of students—some with updates about career changes; a few send pictures of their children.

So I feel a bit of mentoring has gone on. But today's students seem less hungry for education and more interested in financial success. One has to be concerned about the level of interaction in ways never imagined before. I continue to reach a few students and those relationships continue to be special. They encourage me to keep steady on the course. I have children, grandchildren, and "godchildren," from my academic career, and I want the very best experiences for each student that I have encountered.

What career directions have some of your former students taken?

Twitty: They have gone primarily into medicine and dentistry. Perhaps five percent have entered into careers in pharmacy. Very few have entered graduate school.

Why do you think they have not gone to graduate programs in the sciences?

Twitty: I think I understand why. They look at the medical profession or pharmacy and they see possible financial success, presumably without much external interference. They look at us. They are aware of the battles that go on to get that degree, to get that job, to get tenure, with very little comparable financial comfort. What they see does not ensure job security or satisfaction. Too much depends on things over which they appear to have little control. So why should they go there? They find the continued demands out of sync with the rewards.

How did marriage and having a family impact your career?

Twitty: Early on, I arrived at work at 5:30 to 6:00 a.m. By class time at 8:00 a.m, I had one electrophoretic gel running. I prepared lectures and tried to write proposals and research notes after the children were in bed. That left little time for my husband, who, rather than divorce me, joined me and received two master's degrees at the same time I received the doctorate. He was most compassionate and understanding.

From my research, most women in science who stay married have an understanding mate.

Twitty: It's the only way to do it, or have no husband at all. The probability of maintaining a relationship during the grind of graduate school is nil unless you are similarly involved.

How many children did you have?

Twitty: I have twin daughters. Neither one chose to enter a science career, but they are both professionally trained and educated.

What are some of the projects that you researched along the way when time permitted?

Twitty: My first bit of research was for the master's degree in parasitology. I examined the role of *Eimeria tenella*, a protozoan parasite found in chickens. The parasite initially invades the caecal pouches of chickens but multiplies and invades the blood, causing death within a short period of time. The study revealed the devastation the parasite has on various blood cell populations over the course of the infection.

The doctoral work was in biochemical genetics. I examined a number of isozymes of important enzymes found in variant species of mice. The work analyzed the various isozymes electrophoretically in an effort to determine the relationship between the isozyme function in mice and in humans.

My current interest is the phylogeny of tardigrades, which are invertebrates of uncertain phylogenetic status notable for their capacity to carry out
anhydrobiosis. These tiny animals can lose significant amounts of water, become inactive as long as negative environmental conditions persist, and then
rehydrate upon return of favorable environment. Their relationship among
invertebrates has not been determined. My approach is to compare them ge-
netically with several possibly close animal groups through the analysis of
nucleic acids.

I completed a sabbatical leave in the office of Environmental Justice at
the Environmental Protection Agency. I felt that I needed a totally different
challenge. Since that encounter, I have become increasingly more involved in
the environmental justice movement, both from the perspective of an academic discipline as well as a grassroots activity. I am working with a campus group to establish a multidisciplinary environmental studies program
at Howard. I participate regularly in the D.C. Coalition for Environmental
Justice and the Environmental Roundtable. I have presented talks at a number of Brownfield Conferences, which are put on by environmental groups.

Is there an environmental science program at Howard?

Twitty: Not in the sense of an established curriculum. Howard lags behind in recognizing the need to help prepare students in the environmental
arena. Clark Atlanta, Xavier, and Southern Universities and a number of
other minority institution,, not too mention a host of majority institutions,
have solid programs. This is tragic, not only because we are not preparing our
students in an area that negatively impacts our people, but we are missing out
on funding that is available for research. Later this summer, I will participate
in another workshop designed to enhance my computer information base
with respect to some of these issues.

> *It's wonderful that you have found another area that is of vital impor-
> tance to African Americans. How do you think black women can be-
> come more visible in science?*

Twitty: It is important that black women become more visible in science.
The group called Minority Women in Science should have a roster that includes hundreds of minority women rather than the few who have kept it
going for twenty years. We must become more vocal in our respective professional organizations and seek committee assignments and ultimately major offices. It has been said that many African Americans seek medical ca-

reers because they are unaware of other meaningful career choices. It is our responsibility to make them aware of other opportunities and to somehow convince them that while financial security is desirable, there are truly rewarding careers awaiting them.

So what has kept you going all these years?

Twitty: First of all, my upbringing, which demands that I continue to try to make a contribution and that I never forget my roots. My mother instilled in us a very strong sense to give back. Next, my students deserve the best that I can provide. After all, one day some of them will replace me. And of great of importance to me are those students who, after achieving success of their own, return to say thanks. No amount of money or any other recognition can replace the satisfaction that brings. Not too long ago, I had experienced a new low because my students were not performing as I expected. A knock on the door came from a former student. The student had successfully passed residency requirements and was leaving for a research position at Harvard University and wanted to come by just to say thanks. That gesture made my day and renewed my drive to turn the current class around. Teachers are blessed with the most important job in the world, for they have such a great opportunity to influence the lives of so many.

So you may have a few moments where you feel like you are not reaching your students, but one student can make all the difference. Do you have any regrets?

Twitty: No, I really have no regrets. I still consider that I have been afforded and have taken advantage of opportunities as they have been presented. I consider teaching to be the most honorable profession, because it provides the opportunity to influence young minds. It presents a challenge which constantly takes on a diversity of appearances. My life has been enriched by my experiences beyond expectation.

What advice would you like to offer young women considering a career in science?

Twitty: The most important advice I can offer is that she must believe in herself, that she must respect herself, and that she should never allow anyone to limit her capabilities or to diminish her vision of a goal.

Interview Date: December 1996.

Selected Publications

Gaylord, C., and G. W. Twitty. 1994. Protecting endangered communities. Fordham Urban Law Journal.

Twitty, G. W. 1991. Contemporary Zoology: A Modular Approach Text to General Zoology for Entry Level Zoology Majors. Howard University Press.

Carter, J., F. Heppner, R. Saigo, G. W. Twitty, and D. Walker. 1991. The state of the biology major. BioScience, 40, 578–683.

LaVern Whisenton-Davidson

A Passion for Mosquito Research

THAT CONTINUOUS BUZZING NOISE AROUND YOUR EARS is enough to make any human scream. But those annoying bugs have fascinated LaVern Whisenton for the last fifteen years. She has done some outstanding research using the mosquito as a model organism.

A native of St. Louis, Missouri, LaVern grew up with five brothers and one younger sister. She recalls, "My mom had a lot of boys and then my younger sister and I were born in the 1950s. My mom made sure we were raised to be strong, independent women." LaVern credits her mother and high school biology teacher for encouraging her to pursue a career in science.

After high school, LaVern went to Iowa for her undergraduate edu-

LaVern Whisenton-Davidson, a professor of biology at Millersville State University of Pennsylvania, finds an ideal environment for research and teaching.

cation. She began as a freshman at Iowa State University, but because of the high tuition cost, she transferred to Morningside College, a less expensive Methodist college in Sioux City, Iowa. She excelled at Morningside and graduated on the dean's list. From Iowa, she went to the University of Notre Dame, where she received both her master's and doctoral degrees.

With degrees in hand, LaVern was ready for the exciting world of scientific research. She applied for and won two significant postdoctoral fellowships at Notre Dame, where she received training in parasitology and vector biology for three years. Later, she began her work on mosquitoes with Walter Bollenbacher in North Carolina. After a few more years of research training, LaVern was an energetic, well-trained, and highly qualified scientist; but she still had not landed an ideal position. After several interviews, a positive response came from Millersville University in Millersville, Pennsylvania. It was an ideal position for LaVern because it offered a small university setting along with the opportunity to continue her research. After settling into her new position, LaVern received several grants to continue her research, both at Millersville University and the University of North Carolina. She is also regarded as one of the most effective instructors in her department. She teaches and mentors students of all races and genders into productive careers.

Along the way, she met and married a wonderful person later in her career. Her younger sister also followed her scientific path and became a pharmacist. LaVern's research has carried her to international conferences around the world, but these days a relaxing evening at home will do just as well. She puts it best when she says, "it used to be about the science but now it is about being balanced."

How did you become interested in science?

Whisenton-Davidson: I was always interested in biology. When I was in high school, I always liked the biology courses and I took advanced biology and really liked it. When it was time to go to college, I applied to nursing school. I got accepted and said to myself, "Well, I don't necessarily want to work with humans. I don't want to have their lives in my hands." So I started thinking about finding cures for disease and I thought I wanted to do something that would help people of my race. Actually, I started thinking about doing something with sickle cell anemia but I begin working from the mosquito end. You know, sickle cell anemia was the result of a genetic mutation. People heterozygous for the sickle-cell gene are better able to survive malarial infection. In a backwards way, I ended up working with mosquitoes because of my interest in people who had sickle cell anemia.

> *So this interest started back in college. You were an entomology major.*

Whisenton-Davidson: No, I was a biology major. I went to a small college. I had been at Iowa State University, but the tuition was too high because I was an out-of-state student. So I went to a small Methodist college because the people at my church told me about Morningside College in Sioux City, Iowa. That's where I graduated.

> *Was there any particular person or family member who influenced you?*

Whisenton-Davidson: Dr. Veal in high school. He encouraged me to take the advanced biology course. My mother encouraged me, too. She had a lot of boys and she really encouraged my sister and me to go on with our education. She really encouraged us to get some type of position that would allow us to take care of ourselves and not have to depend on someone else to take care of us. My sister is a pharmacist.

> *Where did you grow up?*

Whisenton-Davidson: St. Louis, Missouri.

> *Do you think race or gender has played a role in how you are viewed as a scientist?*

Whisenton-Davidson: That's a hard question. It depends on who's doing the viewing. In terms of other scientists who were in my area, no, it hasn't. In terms of my career and doing the various things I did, sometimes. Like when I was working on my doctorate, I later found out that one of the people on my committee thought that the only reason I was in the program was because of race, that my department was trying to make its quota. When I did really well on my orals and written exam, he was truly shocked. Actually, it had a positive influence. After all this, he told me that was how he felt. He didn't know some other professor had already told me that, anyway. My performance kind of improved his impression of people of color being able to be successful in science. I think it has an effect on students, that is, how you're perceived and received by students.

Many times at the beginning of a course that I lecture in, when the students first see me, they are shocked, because my last name, Whisenton, throws them off. In other words, it doesn't sound like the typical African American last name. At other times when they see that it is a person of color, they will challenge me to really prove myself. As the semester goes on and once you pass their "little test," I guess they forget that you are a person of a different race, and the course proceeds as normal. That's not something that my colleagues have had to experience. They don't have to prove themselves as much.

> *Do you think gender is an issue, and if so, do you think it's difficult to distinguish sometimes between race and gender issues?*

Whisenton-Davidson: I think gender is important, but as a black female I find it very easy to distinguish race and gender issues.

> *You're the first one to say that you can easily distinguish race and gender issues.*

Whisenton-Davidson: White women may not distinguish the differences in the problems that occur. They feel like the woman of color has the same problems, but I think women of color have different issues that we have to address. One of the courses I teach is anatomy and physiology for nursing students. One section was taught by a white woman, and she taught the first semester while the regular professor, a male, was on leave. They killed her in the student evaluations. When the male professor returned, they said, "Oh, he was wonderful. He was so accepting." She was just as good—if not a better—lecturer, I felt, than he was. That's a clear example of a gender issue. When you teach in some of these clinical courses, it becomes an issue. The students are so accustomed to looking up to male authority figures like doctors. Of course, the same is true for the professorate. They are more accepting of a white male teaching a course than they are of a white woman. But the real issue is that the white woman professor does not recognize that the students are even less accepting if it is a black woman professor. The black woman must contend with both race and gender. Some white women just don't seem to get that. And in terms of health professional students (especially nursing students), I believe if they had an order of preference it would be white male, black male, white woman, black woman.

> *Do you think that the civil rights movement played a role in your selection of and success in science?*

Whisenton-Davidson: Oh, yes. I would not have been able to do the things

I've done if the civil rights movement had not been during the time I was growing up. Those issues became very important in my being able to do certain things. When I was in grade school, we were still segregated, and then I went to an all-black high school. As I was coming out of high school that's when the civil rights movement was really becoming active, and it was extremely helpful in terms of getting scholarships for undergraduate and graduate school.

Do you feel the women's movement played a role?

Whisenton-Davidson: Not really. I feel that the civil rights movement played a bigger role to me than the women's movement. I imagine it did have an effect, but for me it was so subtle.

Is there any other issue or obstacle that you think could have affected your success in science?

Whisenton-Davidson: I can only speak of my area. Race does not play an issue in terms of other scientists who are in the same area I'm in. However, when you start looking at the national grants competitions, for example, race in a larger context becomes important when you have people judging you and making decisions about your future to do research and your students' future. For instance, when you look at where the people of color in science are getting teaching positions or where they choose to teach, they're mostly at a predominantly black college or university. Those people—the researchers, panel reviewers, etc., or anyone in a position to make a policy decision—who are judging do have an attitude or a preconceived notion of that person's ability and so before they've even read or judged the research, they may already have an opinion such as, "Oh, they're at that school. They're not doing much." They say it's not an issue but of course it is.

I was wondering about other factors like age.

Whisenton-Davidson: Age may become a factor for a person if she chooses to leave the field for a while, raise her family, and then come back to the science arena. Science is moving at such a pace that any interruption can be a hindrance to keeping abreast. People pay attention to that. That is, sometimes women are judged more harshly than men because they will take the view that she's been away from the field for a long time and they will wonder whether she can reconnect to where the discipline has moved. I think that notion can be a hindrance.

How do marriage and family play a role in your life as a scientist?

Whisenton-Davidson: It's mellowed me out. At first, it was always just the science, but now I've come to realize there's more to life than just the science. A full, balanced life is really important.

> *Well, in your case, it was after tenure that you got married, so do you think women should wait until after tenure?*

Whisenton-Davidson: It is helpful because when you're a person of color they expect you to do more work for tenure. If you look at the people who have gotten tenure and compare it to me, I feel I had to do a little extra. They expected extra time, which they say they don't, but I know that it helped in my case. I think because I was single and I was able to put in that extra time, it helped. When I applied for associate professor, one of the people on the committee said, "Well, you have produced enough for someone going up to full professor." It may not be fair, but you almost have to do that as a person of color. You have to be a super-duper hero. The whole tenure and promotion is a very tight and difficult judging process, anyway. I was able to put that kind of time in when I was not married. I couldn't put that kind of time in now. I wouldn't want to.

> *Because black scientists are rare and we are often viewed as invisible, what can we do to become more visible in science?*

Whisenton-Davidson: The only thing I can think of is to keep on doing and trying to make sure that your science is good and become part of all the organizations related to your field. We need an organization for women scientists of color.

> *There is the National Network of Minority Women Scientists based in Washington D.C. under the AAAS, but it is not very active now.*

Whisenton-Davidson: There may be one, but if it is, it's not very organized or visible. We need one that cuts across scientific disciplines. The only organization that I am active with is the Organization of Black Scientists, which is out of D.C. and it is closest to me. It is a local group, and I found it through Elvira Doman. I think another important thing is some type of mentoring. Dr. Doman has been instrumental in helping me develop my career after I finished my education.

> *You met her as a student?*

Whisenton-Davidson: I met her at a meeting when I was a post-doctoral research associate. She is a representative of the National Science Foundation, and I think that we need those older role models who are supportive. She is very supportive of people of color whom she meets and tries to encourage them to stay active in science. She's kept me going. She recommended me to be on the panel of the National Science Foundation so I could stay active. We need more people out there who have made great strides and who are doing well in their chosen field of scientific endeavor.

How do you envision yourself as a scientist and where do you see yourself in ten years?

Whisenton-Davidson: Because I am at a smaller institution, I see myself spending more and more time being concerned more with the educational aspect, getting women and students of color interested in science, than hardcore research itself. Right now, I'm trying to get a grant for a math and science enhancement program. I see myself trying to seek funds in those areas and less in the research. I do some research, but with my teaching load it's very hard to do. It's not because of race or anything like that but because of the kind of institution I chose to work at. It's changing the direction that I'm going in.

Would you comment specifically on your scientific research?

Whisenton-Davidson: My research involves a study of the hormones, neurohormones, and/or chemical factors involved in regulating molting and metamorphosis in holometabolous insects. Specifically, I am concerned with the processes in mosquitoes. Biochemical, immunological and molecular techniques are used to study these insects.

What advice would you offer young women, particularly young black women, who wish to become scientists?

Whisenton-Davidson: I'd tell them to go ahead and pursue it. There's nothing that should be in the way. If one has the perseverance and is hard-headed enough, one can become successful in any science field. A career in science opens all types of avenues. There are a lot of paths that young women can take, whether it's health care-oriented, science education, or a researcher in a university or a company. There are so many avenues.

What do you say to the young woman who would say to you, "I don't see how you can have both family and career, particularly if you're going the academic route"?

Whisenton-Davidson: I think it's hard to do. I know it can be done because I've seen people be successful in both. For a black female, it's extremely hard to do. After you've reached part of that goal (say being a successful research scientist), then you can make some decisions about family and personal life. I guess I am not the best person to answer this question, because I put science first early on in my career. It's about making life choices. Earlier I chose science instead of marriage, but it's a hard thing to do.

Do you have regrets?

Whisenton-Davidson: No. I'm so glad I waited because that can have its good points too. Marriage and family life can happen, but you might want to push it further down the line in terms of when you want to do it, which can be good. It really is a personal decision, but I think young women should be aware that early bad personal choices can influence future career opportunities. Waiting gave me the time to find the kind of person that I was really compatible with. For me, it was a good thing and I have no regrets that I waited.

Interview Date: May 2000

Selected Publications and Research Activities

Whisenton, L. R. 1980. Biological, biochemical, and hormonal aspects of ovarian development in Toxorhynchites brevipalpis (Theobald). Doctoral dissertation, University of Notre Dame, Notre Dame, IN.

Masler, E. P., L. R. Whisenton, D. A. Schlaeger, S. H. Kang, and M. S. Fuchs. 1983. Chymotrypsin and trypsin levels in adult Aedes atropalus and Toxorhynchites brevipalpis (Theobald). Comparative Biochemistry and Physiology, 75B (3), 435–440.

Bottjer, K., L. R. Whisenton, and P. P. Weinstein. 1984. Ecdysteroid-like substances in Nippostronglyus brasiliensis. Journal of Parasitology, 70, 986–987.

Whisenton, L. R., and W. E. Bollenbacher. 1986. Presence of gonadotrophic and prothoracicotropic factors in pupal and adult heads of mosquitoes. In Insect Neurochemistry and Neurophysiology (A. B. Borkovec and D. B. Gelman, eds.), pp. 339–341. Humana Press, Inc., Clifton, NJ.

Whisenton, L. R., T. J. Kelly, and W. E. Bollenbacher. 1987. Isolation and partial purification of gonadotropic factors in heads of pupal and adult Aedes aegypti. Mol. Cell. Endocrinol., 50, 3–14.

Whisenton, L. R., N. A. Granger, and W. E. Bollenbacher. 1987. A kinetics analysis of brain mediated 20-hydroxyecdysone stimulation of the corpora allata by Manduca sexta. Mol. Cell. Endocrinol., 54,171–178.

Whisenton, L. R., J. Warren, M. K. Manning, and W. E. Bollenbacher. 1989. Ecdysteroid titres during pupal-adult development in Aedes aegypti: Basis for a sexual dimorphism in the rate of development. J. Insect Physio., 35 (1), 67–73.

Nogueira, B. V., L. R. Whisenton, R. S. Gray, and W. E. Bollenbacher. 1996. Life cycle expression of a bombyxin-like neuropeptide in the tobacco hornworm, Manduca sexta. J. Insect Physiol., 43 (1), 47–53.

Research Awards

University of North Carolina-Howard Hughes Program for Minority Advancement in the Biomolecular Sciences, $20,000 (1992); $20,000 (1993); and $10,000 (1994).

National Science Foundation (NSF) research award #DCB-8903769, Steroidogenic neuropeptides in mosquitoes, $124,000 (July 1989–July 1992).

State System Social Equity Office Award, Academic Excellence Program/ Math and Science Initiative for $25,000 (1992–1993).

Pennsylvania State System of Higher Education (SSHE) Faculty Professional Development Council Grant Award, Development of an in vitro assay for studies on neurohomonal regulation in mosquitoes, $4,800 (1989).

EPILOGUE

Continuing to Tell the Story

FOR YEARS I HAVE YEARNED to see our stories told—if not in a book, then in articles; then as I matured I longed to hear the stories of black woman scientists told in their own voices. No one had ever done that to my satisfaction. *Sisters in Science* is an amazing personal and professional victory for me. Many times I have been told to steer clear of "non-scientific writing," but I am a risk taker and a woman who ultimately follows her heart. It is the only way that I have found peace in my life. I did not listen to my critics or well-meaning encouragers when they told me not to do any "soft" writing—meaning human interest stories. But I did it anyway. Thereafter, *Sisters in Science* gave me much sustenance as I waded through all the politics of tenure and life. This book had to be born.

Yet the stories in this book only begin to tell the story. I know that there are as many stories to tell as there are black women scientists, who hail from all walks of life and who excel in various disciplines in science, engineering, and mathematics (SEM). As I listen to the familiar voices I love so dearly in *Sisters in Science*, I see parts of myself and my story reflected in nearly every story I read. Yet there are parts of my story that I do not hear. I know the same is true of all the women, and it is my dream that more of our stories will be told and that more women will actually write their own stories themselves.

Five years ago I found myself at a crossroad. I was certain of only one thing: I had to leave the environment to which I had grown so accustomed. By most accounts, I was a success. I had attained tenure and promotion, was well on my way to full professor—a status that my former chairperson firmly believed I should have already attained at my first tenure and promotion review. My publication list in major national and international journals was extensive. My service record was even longer, a sizable portion of which I was not even allowed to include in my tenure and promotion file. Still, it was an amazing list of service projects completed throughout the university community, the city, the state and even the nation. This extensive record of service did not include all the sacrifices I had made to assist students, especially African American and Asian students, who often needed advice and encouragement late at night and at other inopportune moments when I really should have been attending to my own personal and professional development.

After all my diligence and sacrifice, most people could not understand why I would leave the proverbial ivory tower. "You've made it," they said. "Only a fool would leave a tenured position in a research university and go elsewhere—at least not without having another job in place." I did not listen. Am I ever grateful I didn't! I was finally woman enough to know what I wanted and how I wanted to define myself and my future. Yes, I had learned to play the academic game well; I survived the publish or perish syndrome. I had even learned to thrive in a stultifying environment that had used me in every imaginable diversity ploy and numbers game.

And yes, there were plenty of good times too. I enjoyed my work and loved interacting with new people and exploring new research projects and experiences. But I did pay a price for it; it is difficult to chart new territory, especially when there are no road maps and very few guides who look like you to show you the way. It is often a peculiar predicament that some black women find themselves in when they are in new territory. They are often pioneers in their scientific disciplines. For that reason, there are times when only another African American woman scientist can possibly understand the dilemma in which I found myself. And more often than not, for me it is in her story that my story is most firmly rooted.

This is true even when our experiences do not overlap perfectly. That is why I can boldly say that even as my story takes a different path from many of my sisters in this book, I know these women's voices. I know their joys, I know their pain. I know their triumphs, I know their despair. I know these women and they know me. We share a form of marginalization that others find difficult to grasp and that we often have a hard time expressing in terms that others really understand. Even with well-meaning, sensitive non-African American women scientists, our experiences are simply dismissed or so misunderstood that we give up trying to make others "see" what we really feel. Many times we settle for civility in the workplace and in the academy. At least we can all get along well enough to get our work done. Yet that means the core of the problem remains untouched. Our hearts are the same. We really only get to know each other superficially.

The good news is that if we remain in the same place long enough, we may shed some of the facades and allow some forms of polite professional intimacies to occur. Time does bring about a change, and most people are willing to engage in some politically correct banter over time. As I reflect on my Missouri days, I think one of the most beautiful things that came out of that experience was that my white male colleagues and I finally had to learn to see each other as one human being relating to another. I stayed at the university for over ten years. In time, we began to look beyond each other's race, gender, and class. It is looking beyond those external factors that allowed all of us to break barriers. We all weathered the storm that my presence seemed to have brought to the department and the university. I chose to leave when I was ready on my own terms. Time, experience, and, yes, some personal sacrifice on my part did bring about a drop of change in my former department and the culture of science. Even when it was not a warm environment, I learned a lot about science, how some men approached science, how they saw themselves as professionals, how some women scientists patterned themselves like men scientists, and more importantly, how I really wanted to do science. I took the good and the bad of those experiences to set out on a new path as a scientist, educator, and person. I chose to live by faith and let the best of me emerge to reach my fullest potential. I will not say I had no moments of self-doubt and wanderings, but by and large I knew I had made the right decision.

A few of my sisters' academic preparation and professional careers have developed mainly in historically black colleges and universities, while others have taken place in predominantly white universities, and others in both. I fall into the latter category, with most of my adult life having been experienced in predominantly white environments.

My preparation at Tuskegee University (then Institute) was simply wonderful in terms of its nurturing my scientific curiosity and exposing me to various fields in science. It was not a perfect place, but it was perfect for me. If anything, it protected me from the harsher realities of life even as it prepared me to thrive later. As I scurried about the campus, exploring its labs, and intermittently prepared to become some kind of doctor, I had no idea I was going to enter strange, uncharted waters. If Tuskegee was a sign of what was to come, I would have thought that I would have lived among a thriving class of African American professionals. Even John Andrews Hospital, which sat at the edge of the campus, was filled with African Americans doctors, nurses, nutritionists, and other professionals. My classes had just as many African American women as men. Most of us wanted to be doctors of some type, rather than researchers or university professors, which was probably a reflection of our limited exposure to African Americans in the scientific profession. When I participated in a United Negro College Fund summer program at Fisk University in 1976, I met other African Americans from all over the country who had the same idea as I. Once again, I never felt alone or isolated.

I was thriving among a rising group of aspiring physicians, a sizable number of whom were women. We were so full of life and excitement that we didn't have time for fear. If we feared anything, it was the fear of failing organic chemistry. I had no idea that the Tuskegee cocoon was the first stage in my development and that the world would be a different place.

At Alabama A & M University, I began to see a different world, yet it too was nurturing in its own way. For the first time, I stepped outside the realm of the familiar world of biology onto a path less traveled by women. I was the only woman getting a master's degree in soil microbiology in a class of maybe 10–15 males pursuing degrees in soil or plant science. The only other woman who was enrolled in the program was an undergraduate. Fortunately, we became close friends, and she remains one of my best friends. But for the first time I began to feel a tinge of what it means to be the only female in a class of men. In my first soil science class, there were three females in a class of about fifty men. Actually I performed better at the graduate level, in part because I was more focused and I was beginning to understand the importance of self-discipline and having a vision of what I wanted to achieve. But A & M was still a historically black college. I was still in familiar territory, and while I felt the first stings of professional isolation, I was still among friends in warm, familiar waters.

Neither Tuskegee nor A & M prepared me for the social climate and emotional isolation that awaited me at Michigan State University. Most of my associates would say I didn't seem to have it that hard. Somehow I managed to pass my classes, my qualifying exams, and dissertation proposal on a reasonable schedule. And in some ways I led a very routine life as a Ph.D. candidate, but I also know the internal struggles I faced and the racism I experienced. I also know the onset of what I now call the disease of exclusion and the sometimes self-imposed isolation to cope with that exclusion.

On the surface I appeared to be accepted by most of my student colleagues and most of the faculty, and I must say that I actually was for the most part. I have always been friendly and outgoing. Most people found me easy to get along with, and I made friends easily, but navigating in chilly waters will chill even the friendliest person unless he or she makes an extra effort to offset it. Herein lies the difference: it is putting in all the extra effort that gradually drains one of fuel needed to complete the work.

At Michigan State I found myself swimming in new, cold waters socially and academically. For the first time I found myself at times marginalized even in groups that I had always counted on. I had never before had anyone call into question the depth of my "blackness" just because I was friendly to all people. But that was one of my first encounters with some of my black counterparts at MSU. I had a nasty experience in the graduate dorm cafeteria with a black male who questioned my blackness when I chose to sit on that particular day with a white friend. I had quite a few disconcerting events with some of my fellow black graduate students (male and female). If that was not enough, I was one of only a handful of females that entered my department that year. Most of the classes were dominated by white male students. When study groups were formed, I generally found myself excluded. In one class, when I approached the lone white female student to see if we might study together, she made it clear that she wanted total distance from me. She only came back three years later, when she had experienced failure, to extend the hand of acquaintance. By then, I had a firmly entrenched cadre of friends from all over the world to call upon as friends.

Perhaps the hardest thing to explain to people is how one is judged and her work is judged when she is African American. When I arrived at MSU, for the first time in my life, I found that the person I brought to the table was not enough. Though I would never claim that discrimination, gender issues, and even racism do not exist at black colleges or universities—yes, I did have a number of negative racial assumptions made about me by some of my well-meaning white professors at Tuskegee and A & M—I will affirm that by and large, I rarely had my existence and my right as an American to pursue a degree questioned in the way that it has been since leaving my black universities. The questions are rarely stated so blatantly and they take various subtle forms, yet the very nature of the question, the very asking of the question itself, suggests that I am not perceived as good enough to do a particular job until I have proven myself over and over again. Having to prove myself might be fine if I didn't see inexperienced white men and sometimes white women given opportunities so that they can gain experience and exposure. They don't seem to have to prove themselves a thousandfold to gain opportunities. Opportunities are more often than not their birthright. I'm by no means suggesting that white men or women do not face challenges or that I never received any assistance from white men and others. I am grateful to those white professors and colleagues who showed understanding when I needed it and accepted my shortcomings as well as my strengths. I am simply suggesting that if a black woman enters an environment that is more likely to welcome her presence, there are certain opportunities inherent in that access that will make her life a little easier.

I'm not sure when an acceptable level of proof occurs, but if and when that "conditional acceptance" comes, it opens the doors to all the rights and privileges that inclusion brings. That is one of the reasons that I now look back and say with fondness that isolation truly has its blessings, too. I never felt I had to prove myself to anyone at Tuskegee or at Alabama A & M. I was simply Diann Jordan, a student, who had come to learn and who would leave upon graduation and make everyone proud. For the first time in my life, I began to sense that no matter what I did, for some people in a position to judge my work, it would not be good enough. I would have to prove myself over and over again, a sentiment echoed by some of the women in this book both publicly and privately.

Now I am in the same position I walked into nearly thirty years ago when I saw my very first African American female scientist walking the corridors of Armstrong Hall at Tuskegee Institute. At Alabama State University, I see far more women in my biology class than young men. Even though I teach primarily non-majors, I often have the opportunity to touch the lives of aspiring physicians, health administrators, and, hopefully, a few researchers. When I look at the roster of biology and chemistry majors at Alabama State University, I see a plethora of young women, their numbers exceeding the number of young men. While I am encouraged that more young women are pursuing science—even though most are still not interested in research careers—I am not fooled. I know that by the time these science majors could potentially earn a Ph.D., the number of women doctorates will be miniscule compared to their male counterparts. They will earn degrees in fields of medicine, but most researchers will be young men except in fields like biology. The pool of young women entering the professional research science, engineering, agricultural sciences, and mathematics pipeline is still small.

The question remains why. Are African American women that afraid of science and mathematics, even when they are willing to dip their toes into these waters at the undergraduate level? I don't think so. Enough of them graduate with the bachelor of science degrees, so that can't be true. The data presented in this book along with data from the National Science Foundation and current studies at HBCUs show that this is not the case. Why do young women choose not to continue in that science pipeline to pursue doctorates and, perhaps, research careers? I think that the young women who might pursue doctorates in SEM are very savvy about their future. They look at the long hours in laboratories, they look at the social sacrifices such paths often take and sometimes the emotional drain they will incur. If that is not enough, they compare what they could earn as physicians and pharmacists or health administrators with what university professors earn and the answer is clear. The payoff is just not there, monetarily or in any other way. They decide to say "no thank you" to research and find a more amenable, less intrusive path to their practice of science. As much as I would like to tell these young women things are changing rapidly in the world of science and that they will be accepted, I have to be realistic at the same time. A science career is not for everyone. But for those young women who truly want a career in science I think we owe it to them to continue to work diligently to open up the pipeline and give them the best possible advice that we can.

Having told this side of my story in science, I will say that men played a major role in my becoming a research scientist. They were supportive at key stages in my career, which helped me become successful in science. That part of my story began in junior high when my male science teacher first introduced me to the world of science and my eighth-grade math teacher, who happened to be a white male, taught mathematics in such a clear and concise manner that I loved mathematics. But it was my high-school biology teacher who really opened my eyes to science and told me that I could be anything I wanted to be. That story continued in college with both black male mentors and black women role models. And later, I know, it is only because of the vision and kindness of men like Robert Taylor (who helped to redirect my whole graduate education by advising me to go to Michigan State), Jim Tiedje (my Ph.D. advisor), Charles Rice, Greg McDonald, KC Morrison, Ernie Kung (former department chairperson), Al Vogt, Bob Kremer, and others who really made a difference in whether or not I would have a successful academic science career. Later, powerful women scientists like Judy Wall, a biochemist, were very influential in my wanting to continue the research part of my career, and she provided key opportunities for me to grow professionally. But there are not enough of them who will genuinely put forth the effort. That is a part of the problem. Obviously, white males mentor minority students, but many do not see it in their best interest to train the next generation. Just as they have been a part of the problem, they now must be a part of the solution. The African American student will bring a whole history, culture, and attitude to science that some in the majority race or gender will be unaccustomed to. It is important that those differences are embraced, understood, and even expanded by science and engineering advisors to help develop the best scientist possible. This is sometimes done in what seem to be very small ways. I can remember my advisor insisting that I attend a laboratory cross-country ski trip in northern Michigan. I was an Alabama native, but he wanted to make sure that I experienced something a little different from my own comfort zone. It was very cold, but I remember it as a positive experience. Some minority students will be well-prepared for graduate education and others will not. Advisors have to make sure those students who have a great deal of potential but need a little more time and nurturing do not fall through the cracks. As some of these women scientists have stated, "we often have to meet the students where we find them." The face of science is changing, and white males and others have a stake in making the transition smooth. We all have a role to play in changing the face of science by increasing the number of women and minorities and providing opportunities for anyone who is different from the status quo in the scientific and engineering pipeline.

Most women who will earn doctorates in SEM will receive their education at predominantly white institutions. I love science and I would love to see more young men and women pursue careers in SEM. My prayer is that my younger sisters and brothers of all races in science will write a different chapter in the unfolding saga of how we do science. I know the days of "the first African American female" sadly enough are not over; yet, the story of being first does not have to be tainted by isolation and discrimination.

African American women scientists want the same things all other scientists want: to be allowed to pursue and practice their science as freely and inventively as their imaginations and intellects will allow. On a human level, they want to belong and receive fair treatment in their dealings with others. Above all, they want the same dignity and respect that any whole healthy person desires.

The future is bright for those who love science. I still say that a young African American woman should pursue her dream of doing science. I only want her to know that she is standing on the shoulders of strong women who have gone before her to pave a new road for her and for those who will come after her. If she ever needs that kind of affirmation, she can look to these women and know that she is not alone. She, too, will have her story to tell. All of the work in science is important, whether she decides to be an administrator, writer, health professional, policy maker, teacher, or bench scientist. We all have a destiny to live and a story to share. For that reason, I am proud that I did all I could to change the face of science in the place where I found myself. If my presence helped to change the face of scientists in the minds of students and faculty alike, then I am both grateful and vindicated.

References and Recommended Reading

Adams, Eugene. *The Legacy: A History of the Tuskegee University School of Veterinary Medicine.* Tuskeegee, Ala.: Media Center Press, 1995.

Ambrose, Susan, et al. *Journeys of Women in Science and Engineering: No Universal Constants.* Philadelphia: Temple University Press, 1997.

Benjamin, Lois, ed. *Black Women in the Academy: Promises and Perils.* Gainesville: University Press of Florida, 1997.

Bernstein, A., and J. Cock. "A Troubling Picture of Gender Equity." *The Chronicle of Higher Education* B (June 15, 1994) 1–3.

Bernstein, Leonard, Alan Winkler, and Linda Zierdt-Warshaw. *African and African American Women of Science: Biographies, Experiments, and Hands-on Activities.* Maywood, N.J.: The Peoples Publishing Group, Inc., 1998.

Brown, M. "Black Women in American Agriculture." *Agricultural History* 50 (1975): 202–211.

Brown, S. V. "Testing the Double Bind Hypothesis: Faculty Recommendations of Minority Women Fellowship Applicants." *Journal of Women and Minorities in Science and Engineering* 2 (1995): 207–233.

Clark, J. V. "Black Women in Science: Implications for Improved Participation." *Journal of College Science Teaching* 17, no. 5 (1988): 348–352.

Clewell, B. C., and B. Anderson. "African Americans in Higher Education: An Issue of Access." *Humboldt Journal of Social Relations* 21 (1995): 55–79.

Cole, Johnnetta B., and Beverly Guy-Sheftall. *Gender Talk: The Struggle for Women's Equality in African American Communities.* New York: One World/Ballantine Books, 2003.

Cooper, Anna. J. *A Voice of the South.* Xenia, Ohio: Aldine Publishing House, 1892.

Culotta, E., and A. Gibbons. "Minorities in Science: The Pipeline Problem." *Science* 258 (1992): 1175–1237.

Davis, Marianna. *Contributions of Black Women to America: Civil Rights, Politics and Government, Education, Medicine, Sciences.* Columbia, S. C.: Kenday Press, 1982.

Ellis, E. M. "The Impact of Race and Gender on Graduate School Socialization, Satisfaction with Doctoral Study, and Commitment to Degree Completion." *Western Journal of Black Studies* 25 (2001): 30–35.

Essien, F. "Black Women in the Sciences: Challenges along the Pipeline and in the Academy." In *Black Women in the Academy: Promises and Perils*, ed. Lois Benjamin. Gainsville: University Press of Florida, 1999.

Giddings, Paula. *When and Where I Enter: The Impact of Black Women on Race and Sex in America*. Toronto: Bantam Books, 1984.

Harding, Sandra G. *The "Racial" Economy of Science: Toward a Democratic Future*. Bloomington: Indiana University Press, 1993.

Heinemann, S. *Timelines of American Women's History*. New York: The Berkeley Publishing Group, 1996.

Hine, Darlene Clark. "Co-laborer in the Work of the Lord: Nineteenth Century Black Women Physicians." In *The Racial Economy of Science*, ed. Sandra Harding. Bloomington: Indiana University Press, 1993.

Hine, Darlene Clark, Elsa Barkley Brown, and Rosalyn Tergborg-Penn. *Black Women in America: An Historical Encyclopedia*. Bloomington: Indiana University Press, 1993.

Hine, Darlene Clark, and Kathleen Thompson. *A Shining Thread of Hope: The History of Black Women in America*. New York: Broadway Books, 1998.

Jay, James. *Negroes in Science: Natural Science Doctorates, 1876–1969*. Detroit: Belamp Publishing, 1971.

Jemison, Mae Carol. *Find Where the Wind Goes: Moments from My Life*. New York: Scholastic, 2001.

Jordan, Diann. "Historical and Current Perspectives: Black Women in Mathematics and Engineering." *Society of Women Engineers* (March/April 1997): 14–18.

Jordan, Diann, and Jane Ford-Logan. "Black Women and the Workforce in Agriculture and Natural Resources." *Women and Natural Resources* 5 (1994): 4–7.

Jordan, S. M. *Broken Silences: Interviews with Black and White Women Writers*. New Brunswick: Rutgers University Press, 1993.

Journal of Women and Minorities in Science and Engineering. New York: Begell House.

Kessler, James H., et al. *Distinguished African American Scientists of the 20th Century*. Phoenix, Ariz.: Oryx Press, 1996.

Leggon, C. B., and W. Pearson, Jr. "The Baccalaureate Origins of African American Female Ph.D. Scientists." *Journal of Women and Minorities in Science and Engineering* 3, no. 4 (1997): 213–234.

Manning, K. *Ernest Everett Just: The Role of Foundation Support for Black Scientists 1920–29*. Oxford: Oxford University Press, 1983.

Moses, Y. T. *Black Women in Academe: Issues and Strategies*. Washington, D.C.: Association of American Colleges, 1989.

Phillips, M. C. "Tenure Trap: Number of Obstacles Stand in the Way for Tenure for Women." *Black Issues in Higher Education* 10 (1993): 42–44.

Reynolds, Betty, and Jill Tietjen. *Setting the Record Straight: The History and Evolution of Women's Professional Achievement in Engineering*. Denver: White Apple Press, 2001.

Rossiter, M. W. *Women Scientists in America: Struggles and Strategies to 1940*. Baltimore, Md.: Johns Hopkins University Press, 1982.

Rossiter, M. W. *Women Scientists in America: Before Affirmative Action 1940–1972*. Baltimore, Md.: Johns Hopkins University Press, 1995.

SAGE, A Scholarly Journal on Black Women. Special issue, Black Women in the Sciences 6 (1989).

Sertima, Ivan Van. *Blacks in Science: Ancient and Modern*. New Brunswick: Transaction Books, 1991.

Sluby, P. C. "Black Women and Inventions." *SAGE: A Scholarly Journal on Black Women* 6 (1989): 33–35.

Smith, Jessie C., Casper LeRoy Jordan, and Robert L. Johns. *Black Firsts: 2,000 Years of Extraordinary Achievement*. Detroit: Visible Ink Press, 1994.

Sullivan, Otha Richard. *Black Stars: African American Women Scientists and Inventors*. New York: John Wiley and Sons, Inc., 2002.

Takaki, Ronald. *A Different Mirror: A History of Multicultural America*. New York: Little, Brown, and Co., 1993.

Thomas, V. L. "Black Women Engineers and Technologists." *SAGE: A Scholarly Journal on Black Women* 6 (1989): 24–32.

Trent, W., and J. Hill. "The Contributions of Historically Black Colleges and Universities to the Production of African American Scientists and Engineers." In *Who Will Do Science? Educating the Next Generation*, ed. Willie Pearson, Jr. and Alan Fechter. Baltimore, Md.: Johns Hopkins University Press, 1994.

Warren, Wini. *Black Women Scientists in the United States*. Bloomington: Indiana University Press, 1999.

White, Deborah Gray. *Too Heavy a Load: Black Women in Defense of Themselves 1894–19994*. New York: W. W. Norton and Co., 1999.

Williams, Clarence. *Technology and the Dream: Reflections on the Black Experience at MIT, 1941–1999*. Cambridge, Mass.: MIT Press, 2001.

Williams, Reva K. Astronomers of the African Diaspora. www.math.buffalo.edu/mad/physics/williams_reva.html.

Yount, L. *Twentieth-Century Women Scientists*. New York: Facts on File, 1996.